The Shoot Apex and
Leaf Growth

Arthur H. R. Petrie

The Shoot Apex and Leaf Growth

A Study in Quantitative Biology

R. F. WILLIAMS

Division of Plant Industry
C.S.I.R.O., Canberra

CAMBRIDGE UNIVERSITY PRESS

Published by the Syndics of the Cambridge University Press
Bentley House, 200 Euston Road, London NW1 2DB
American Branch: 32 East 57th Street, New York, N.Y.10022

© Cambridge University Press 1975

Library of Congress Catalogue Card Number: 74–21716

ISBN: 0 521 204534

First published 1974

Printed in Great Britain
at the University Printing House, Cambridge
(Euan Phillips, University Printer)

Contents

Acknowledgements		*page* vii
1	Introductory	1
2	The quantitative description of growth	9
3	Phyllotaxis	27
	3.1 Spiral and other systems of phyllotaxis	28
	3.2 The parameters of phyllotaxis	35
	3.3 The Fibonacci angle – irrational or inevitable?	42
4	Shoot-apical systems	56
	4.1 Flax, *Linum usitatissimum* L.	56
	4.2 Tobacco, *Nicotiana tabacum* L.	81
	4.3 Cauliflower, *Brassica oleracea* L.	91
	4.4 Blue lupin, *Lupinus angustifolius* L.	97
	4.5 Subterranean clover, *Trifolium subterraneum* L.	100
	4.6 Eucalyptus, *Eucalyptus grandis* Hill and Maiden, and *Eucalyptus bicostata* Maiden, Blakely and Simmonds	116
	4.7 Wheat, *Triticum aestivum* L.	131
	4.8 Fig, *Ficus elastica* Roxb. ex Hornem.	146
	4.9 Yellow serradella, *Ornithopus compressus* L. *Dianella* sp. (Liliaceae) Narrow-leaf wattle, *Acacia mucronata* Willd. ex H. Wendl Sunflower, *Helianthus annuus* L.	157
5	The dynamics of leaf growth	163
	5.1 Subterranean clover	163
	5.2 Wheat	172
	5.3 Relative rates of change, R and G	177
6	The growth of an inflorescence	183
7	The growth of wheat tillers	199
8	Plant growth as integration	207
	8.1 Physical constraint as a determinant of growth rate	207

Contents

   8.2 Constraint and the genesis of form   *page* 212
   8.3 Organization of the shoot apex   214
Appendix   223
   A.1 Three-dimensional reconstruction   223
   A.2 Volume estimation by serial reconstruction   227
   A.3 Phyllotaxis   230
   A.4 Age equivalence and covariance   236
   A.5 Data processing and presentation   239
   A.6 Cell counting   243
   A.7 Conversion table   244
References   245
Indexes   251

Acknowledgements

The author is happy to record his indebtedness to immediate colleagues of the Division of Plant Industry, and to others who have contributed in many ways to the preparation of this book. My personal assistant Miss R. A. Metcalf is especially deserving of my thanks, not only for dedicated technical assistance over the last five years, but for the preparation of nearly all the text figures in this volume. Her predecessor, too, Mrs T. C. Wallace is deserving of my warmest thanks. My colleague, Dr A. H. G. C. Rijven, ever my sternest critic, has helped more than he is aware by his unfailing enthusiasm and interest. He and Professor F. L. Milthorpe made many valued suggestions for the improvement of the text.

In matters mathematical and statistical I owe much to Mr G. A. McIntyre, and more recently to Mr W. J. Muller of the Division of Mathematical Statistics. They do not necessarily concur in all the procedures that have been adopted by the author.

Mr C. J. Totterdell and Mr H. G. Baas Becking, leaders of the Photography and Illustration Sections of the Division have been generous with help and advice, and Mrs A. J. Simpson and Mr E. Brunoro have done all the routine photography relating to the text figures, both half-tone and line drawings. In particular, Mrs S. C. McIntosh and Mrs Joan Simpson are warmly thanked for the dust cover design.

Mrs V. J. Ronning and several members of the Divisional Typing Pool have been very patient with my reiterated requests for typing, and I thank them.

Miss Fumei Lin was involved for a time, as a post-graduate student, in the study of *Eucalyptus* seedling growth. I acknowledge her work and regret that she was unable to complete it under my guidance.

I am especially grateful to the Australian journals of scientific research for permission to reproduce seven plates and 38 text figures previously published by them. They are specifically acknowledged in the legends and References. A few other diagrams have been modified from the originals in the literature, and due acknowledgement is made.

In conclusion, I wish to thank my wife, Ethel, for her forbearance and support throughout the project.

1 Introductory

'. . . if arithmetic, mensuration, and weighing be taken away from any art, that which remains will not be much'.

PLATO, Philebus (Jowett's translation, iv, 104 (1875))

'. . . the growth and development of an organism is the result of a number of ontogenetic processes, among which complex interrelations exist. The interpretation of these interrelations, and of the manner in which the processes are integrated to produce the living plant, is the fundamental problem in the study of growth'.

A. H. K. PETRIE (1937)

At the time of his death, Petrie had completed early drafts of two or three chapters of a book which was to have been called 'The Developmental Physiology of Plants', and the above quotation from Plato had been placed at the beginning of Chapter 2, 'The Change in Dry Weight and Leaf Area, and First Steps in the Analysis of Growth Rate'. Unfortunately the book had not reached a stage from which it could have been completed by any of his colleagues, and we had to content ourselves with placing the quotation on the title page of a bound volume of Petrie's papers in plant physiology for the Library of the Waite Institute, Adelaide.

The second quotation is from one of Petrie's published papers, and epitomizes his thinking and general approach to the study of growth, though more extended statements along the same lines had in fact appeared a year earlier (Ballard and Petrie, 1936). Although his research ouput was quite remarkable in quantity and quality, Petrie did not live long enough to bring this kind of thinking to its full fruition, but he did succeed in transmitting some of his enthusiasm and outlook to his rather few research students and those who became his junior colleagues. This book is intended as a very belated tribute from one who was privileged to be associated with him for almost nine years.

Petrie was the son of Dr James M. Petrie, who was for many years Macleay Research Fellow of the Linnean Society of New South Wales. After holding junior academic posts in Sydney and Melbourne he was awarded an 1851 Exhibition Scholarship, and this enabled him to study under G. E. Briggs of Cambridge. Petrie actually took his current interest in ionic absorption to Cambridge and, with Briggs, eventually published pioneer studies on the application of the Donnan equilibrium to the ionic relations of plant tissues. He was also strongly influenced by the course of lectures on growth given by Briggs at that time.

Petrie returned to Melbourne in 1929, and in 1931 he was invited to join the staff of the University of Adelaide as plant physiologist of the Waite Agricultural Research Institute. He accepted the challenge, and, although he did not immediately give up his interest in ionic relations, he turned his attention more and more to growth and development. In this he was undoubtedly influenced by A. E. V. Richardson, the then director, with whom, in spite of marked differences in temperament, he shared common objectives. One of these was the raising of agricultural research from the level of pure empiricism to one which could boast a sound body of knowledge and theory. Indeed Petrie was one of the first in Australia to attempt to interpret relevant elements of agricultural practice in terms of plant physiological processes.

Petrie believed that the temporal drifts in size, structure and chemical composition of a plant during its development were the results of drifts in metabolism, and with remarkable singleness of purpose he set himself to study these progressive physiological changes. Petrie's usual approach was to alter the external nutrient supply, to apply temporary periods of drought, or to prevent inflorescence development; all as means to the understanding of growth. He saw the necessity for precise quantitative description, and he knew that a great amount of it would be needed before general principles could emerge and be evaluated. To a degree, Petrie's work was still in the data-collecting stage at his death, but it was far more than an elaborate exercise in classical growth analysis. The second quotation at the head of this chapter amply attests a recognition of the significance of organization at the biological level. In particular, Petrie stressed the importance of competitive demand within the plant for metabolites and nutrients. He and others from the group have contributed substantially to our knowledge of the uptake and redistribution of mineral elements during development, and this contribution has been reviewed in some detail (Williams, 1955).

In sum, I believe it is true to say that Petrie's contribution is a large and continuing one; that a considerable body of research in crop physiology and plant response stems from his influence and insights; and that such studies have become academically respectable because of him. Further evidence, if that be needed, is that his voluminous data on the growth of the tobacco plant (collected almost 40 years ago) have recently been made the basis of a simulation study which explores the carbon economy of the tobacco plant (Hackett, 1973).

However, the present book has to do, not so much with the integration of physiological processes as with the prior need for quantitative description. It was this no doubt which prompted Petrie's use of the quotation from the *Philebus*, though, in one sense, this is curiously inappropriate to the main body of biological knowledge. 'Arithmetic, mensuration and weighing' are, of course, tremendously important to genetics, biochemistry and biophysics, but the rest of biology has been, and still is predominantly qualitative in content. Yet one can scarcely say that 'that which remains' – and this includes systematics, morphology, anatomy and the whole corpus of evolutionary theory – 'will not be much'.

Why, then, has so little attention been paid to quantity and time in biology? Some reasons which spring to mind are purely technical. Organisms are variable entities, and measurements, especially growth measurements, are subject to inherent variation which may be many times greater than the experimental errors of actual measurement. This calls for appropriate design of experiments and statistical treatment, and biologists have not always been good at these things. Then, too, many of the critical events in developmental biology take place when the organisms or their parts are exceedingly small. It is easy to look on and record such events with the aid of microscopy, but to record them in terms of quantity and rate is another matter. It adds at least another dimension to the problems of natural variation already referred to.

More potent reasons, perhaps, have been the sheer magnitude of the descriptive task and the limited number of workers. And those that have been attracted to biology have rarely been competent in mathematics or the physical sciences. However, the climate of scientific opinion has also been against the growth of a sound quantitative biology. By that I mean that it has not been fashionable to think of organisms as systems having a hierarchic order which demands investigation at all levels. Weiss (1969) puts the problem very clearly in his chapter in *Beyond Reductionism*. He says, 'There is an age-old controversy in biology between the two opposite extremes of "reductionism" and "holism". The former finds concurrently its most outspoken advocates in the field of so-called "molecular biology". If this term is used to imply no more than a deliberate self-limitation of viewpoint and research to molecular interactions in living systems, it is not only per-

3

tinent and legitimate, but has to its credit some of the most spectacular advances in modern biology. If on the other hand, flushed by success, it were to assume the attitude of a benevolent absolutism, claiming a monopoly for the explanation of all phenomena in living systems, and indeed were issuing injunctions against the use of other than molecular principles in the description of biological systems, this would obviously show a lack of practical experience with, or disregard of, the evidence of supra-molecular order in living systems.'

This incisive statement is well summed up by the aphorism, 'Thought is abstract: and the intolerant use of abstraction is the major vice of the intellect' (Whitehead, 1933). And abstraction is used intolerantly when that which is abstracted from is regarded as unimportant.

I am not qualified to pursue this controversy in depth, nor is this the place to do so. However, it takes very little practical experience to enable one to agree wholeheartedly with Weiss's further statement that 'the *principle of hierarchic order* in living nature reveals itself as a demonstrable descriptive fact regardless of the philosophical connotations that it may carry'. Those wishing to read further on these matters should consult the much neglected book, *Biological Principles* (Woodger, 1929), and the more recent *Hierarchy Theory* (Pattee, 1973).

The central purpose of this book, then, is to provide precise quantitative descriptions of shoot-apical systems of very diverse types. Except for wheat, for which a description of the developing inflorescence is also provided, these are all for vegetative apices. These descriptions were made possible by a development of the old technique of serial reconstruction, which permits the early growth of leaf primordia and related tissues to be measured as volume.

Wilhelm His (1888) drew attention to the importance of measurement for the understanding of morphogenetic processes, and it is to him that we owe the procedure of serial reconstruction. In one place he says: 'The ways of determining the forms and volumes of germs and embryos are somewhat longer and more tiresome than the simple inspection of stained sections; but the general scientific methods of measuring, of weighing, or of determining volumes cannot be neglected in embryological work, if it is to have a solid foundation of facts, for morphologists have not the privilege of walking in easier or more direct paths than workers in other branches of natural science.'

While serial reconstruction has been used many times for the description of form changes in embryos and embryonic organs in animals and plants, it does not seem to have been used in any precise quantitative sense. The procedure is described in the appendix, together with sampling and related procedures which have been found helpful.

Only the studies with wheat (Williams, 1960, 1966*a*; Williams and Rijven, 1965; Williams and Williams, 1968) and subterranean clover (Williams and Bouma, 1970; Williams and Rijven, 1970) have appeared as research papers; the others are now published for the first time. Indeed, it was the need to have all the studies together for ready comparison which prompted their presentation in book form. This has the further advantage that, since all the studies of Chapter 4 have used the same methods, these could be relegated to the appendix, thus avoiding to some extent the clutter of the standard research paper. Even more important was the need for an appropriate medium for the development of the thesis that plant growth is subject to a variety of constraints which need to be recognized alongside those determinants which are accepted almost without question.

I did not become aware of the possible importance of physical constraint as a determinant of growth until the long period of exponential growth sustained by the clover leaf primordium was recognized as an optimal solution to a developmental problem, a solution which involved the system as a whole, and not only the concurrent intracellular events. This awareness quickly led to the recognition that the curious pattern of primordium growth in wheat was also readily interpreted in terms of physical constraint, and this concept began to enter into my thinking when it came to selecting apices for detailed study.

Now the cause-effect relations with which we are most familiar are ones which relate to systems which can be isolated sufficiently to permit the testing of hypotheses *in vitro*. Increasingly, such hypotheses are being checked by in-vivo studies using labelling and other techniques. However, it is suggested that hypotheses relating to the higher levels of organization may be difficult if not impossible to test by experiment. In such cases conviction may have to come, in part inductively, by the comparative study of relevant biological systems, in part deductively, by setting up a general theory which has powers of prediction. Such after all, is the history of the study of evolution and the general theory of natural selection (Huxley, 1954). In what follows, then, I make no apology for drawing attention to the likely operation of physical constraint within the point-by-point descriptions of Chapter 4. To do otherwise would be cumbersome in the extreme. This hypothesis already has experimental backing in that we know that root growth is very sensitive to constraining pressures, and that the early growth and emergence of wheat tillers are subject to physical constraint (see Chapter 7). For all that, only comparative study of developmental systems will show us the extent of its operation, and I suspect that a consistent body of observational fact will remain the best support for possible theoretical developments for some time to come.

[A useful review has been supplied by Vidaver (1972) on the effects of pressure on the metabolic processes of plants. The work reviewed relates mainly to effects of variation in hydrostatic pressure, and probably has little relevance to the localized operation of physical constraint.]

It is a sobering thought that a physicist should have supplied what appears to me to be a quite profound statement on the problem of biological hierarchy (Pattee, 1970). In an introductory statement he says, 'if you ask what is the secret of life, you will not impress most physicists by telling them what they already believe – that all the molecules in a cell obey all the laws of physics and chemistry. The real mystery. . .is in the origin of the highly unlikely and somewhat arbitrary constraints which harness these laws to perform specific and reliable functions. This is the problem of hierarchical control'. Later he gives the more succinct statement, 'If there is to be any theory of general biology, it must explain the origin and operation (including the reliability and persistence) of the hierarchical constraints which harness matter to perform coherent functions'. He warns that such a theory is not simply a set of descriptions at each level, but must concern itself with the interfaces between the levels. The natural tendency to concentrate on one level of organization at a time carries with it the likelihood that the technical languages at each level will become incompatible.

This book, then, is an attempt by a practising biologist to work out the implications of this sort of thinking. It is concerned with the relations of organs within well defined biological systems (shoot apices), and considerable effort has gone into their precise quantitative description. At the same time, the genesis of form is kept constantly before the reader by the use of three-dimensional, scale drawings, and with photomicrographs. The drawings, in particular, have helped greatly in identifying events which were seen to be correlated with changes in relative growth rates of various primordia. Whatever may come of the interpretations which have been placed on those events, there will remain a considerable body of new information about shoot-apical systems. Nevertheless, the notion that constraint is an important determinant of growth rate is not new, as is attested by the quotation from

Richards at the head of Chapter 8. It is also the simplest interpretation of the range of rates set out in Table 2.1 (p. 13).

From time to time and especially in Chapter 8, attention is drawn to the possibility that certain events constitute optimal solutions to specific developmental problems. In particular, this applies to strict exponential growth within systems in which sequences of like members remain in close contact. There are also many problems in morphogenesis, including those of phyllotaxis, to which optimality principles would seem to apply. The reader is referred to Rosen (1967) for an introductory account of the mathematical techniques and some applications.

A secondary theme of the book is that of phyllotaxis. I have attempted to set out and to apply the procedures, which we owe to Richards (1951), for the objective description of shoot-apical systems. To date, these have not been used in any systematic way, and one suspects that most botanists have been daunted by the theoretical detail of the original. Simple geometrical modelling has also been used to study the generation of Fibonacci and other spiral systems from the decussate condition of the dicotyledonous seedling.

Chapter 5, on the dynamics of leaf growth, extends the quantitative description to chemical change in two contrasting leaf types, clover and wheat. To that extent it takes a look at the integration of physiological processes in the growth of the leaf, and it points the way to precise in-vivo study of metabolism in leaf primordia and other embryonic organs during rapid growth. Key concepts here are the relative growth rate R, and G, the rate of production of one metabolite per unit of another based on terminal values for an interval.

The sixth chapter shows that quite complex biological systems, such as inflorescences can be described with precision if the need should arise.

When this book was being planned the intention was to include a critical survey of the concepts of growth analysis, its current status and limitations. The need for this is much reduced as a result of the appearance of *The Quantitative Analysis of Plant Growth* (Evans, 1972), and what little remains to be said would be difficult to place here.

Some other general texts that can be recommended for related reading are, *Apical Meristems* (Clowes, 1961), *Shoot Organization in Vascular Plants* (Dormer, 1972), *The Growth of Leaves* (Milthorpe, 1956), *Growth and Organization in Plants* (Steward, 1968), *The Control of Growth and*

Differentiation in Plants (Wareing and Phillips, 1970) and 'Growth as a general process' (Whaley, 1961). A recent book, *Analysis of Leaf Development* (Maksymowych, 1973) is perhaps closest of all to the subject matter of the present volume. However, it is based largely on one species, *Xanthium*, and the emphasis is upon the phase of leaf expansion, which I have tended to neglect.

2 The quantitative description of growth

'If the rate of assimilation per unit area of leaf surface and the rate of respiration remain constant, and the size of the leaf system bears a constant relation to the dry weight of the whole plant, then the rate of production of new material, as measured by the dry weight, will be proportional to the size of the plant, i.e. the plant in its increase in dry weight will follow the compound interest law.'

'The rate of interest, r, may be termed the *efficiency index* of dry weight production.'

'It is clear. . .that the efficiency of the plant is greatest at first and then falls somewhat, but the fall is only slight until the formation of the inflorescence, when there is a marked diminution in the efficiency index.'

V. H. BLACKMAN (1919)

Although it is commonly acknowledged that we owe to Blackman the first clear statement of the mathematical principles underlying the law of exponential growth, his 'efficiency index' has had an extraordinary history of criticism and rejection. The efficiency index is, of course, none other than the relative growth rate – a concept which has always been eminently respectable. No doubt the nature of the analogy – that of compound interest – and reference to it as a physiological constant are the bases of the misunderstandings. His earliest critics were Kidd *et al.* (1920) and they were effectively answered by Blackman (1920) in the same volume of the *New Phytologist*. Of special interest is the contention of Kidd *et al.* that the *only* way in which plants can be compared is by the comparison of the whole series of efficiency indices throughout their life-cycles. This was a valuable suggestion, even though Blackman correctly countered that, in the absence of such detailed information, the comparison of average efficiency for longer periods is of value. It is certainly not the *only* way, but a very large number of growth studies involving treatment and other comparisons over extended portions of life-cycles have demonstrated its importance.

That Blackman is still being misunderstood will be apparent from a perusal of an otherwise excellent chapter on quantitative interpretations of growth by Steward (1968, see pp. 417–8), and Dormer (1972) says in one place, 'This consideration is sufficient to expose the fundamental

9

artificiality of the compound interest scheme; it represents the growth of a body which remains meristematic throughout.' Macdowall (1972), on the other hand, goes too far in his defence of Blackman's 'kinetically proper expression of rate of plant growth', when he asserts that 'the strange and uninterpretable changes that have been reported for relative growth rate, even in the initial phase of growth, have encouraged suggestions of an elusive "internal factor" and have forced continued reliance on the "technique of growth analysis"'. One can agree that classical growth analysis has put so much stress on the components, net assimilation rate and leaf area ratio, that the relative growth rate itself has been divested of much of its significance. One can also approve the assumption that early vegetative growth is near enough to exponential for Macdowall's type of kinetic study, but not his dismissal of time trends on the supposition that they are uninterpretable. Growth analysis itself has provided partial interpretation of such trends, and internal factors are not less real for being imperfectly understood.

Relative growth rate is so well established as a concept that we would do well to retain the name. Macdowall's proposal to replace it by 'growth coefficient' has little to commend it, though his defence of Blackman serves as a reminder that relative growth rate is in fact the fundamental measure of organic growth. Indeed, it is a superbly sensitive yard-stick for growth not only of whole plants, but also of organ assemblages and individual parts. Fig. 2.1 attempts to justify this claim, and is derived from data presented more fully below.

The three curves of absolute weight change at the top of Fig. 2.1 all cover some six logarithmic cycles of size, so it is scarcely surprising that they conceal more than they reveal about early stages of growth. The same data on a logarithmic scale gives equal weight to all stages because growth, after all, is multiplicative rather than additive in character, except in rather exceptional circumstances. There are some obvious differences between the three sets of data, including the positioning of points of inflexion and the existence of strictly linear segments at different stages for clover and wheat.

Before proceeding further it will be as well to define the concept of relative growth rate. For any attribute of size, W which is changing with time, t the relative growth rate, R at any instant is

$$R = \frac{1}{W}\frac{\mathrm{d}W}{\mathrm{d}t} = \frac{\mathrm{d}\ln W}{\mathrm{d}t},$$

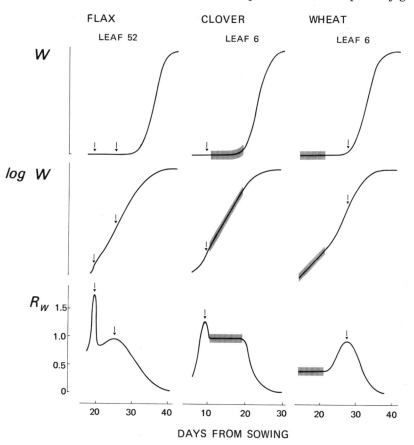

Fig. 2.1. Comparative analysis of the growth curves for representative leaves of flax, clover and wheat. These are given first as absolute weight, W; then on a logarithmic scale, log W; and finally as the relative growth rate (weight-basis), R_W. Maxima for R_W are indicated by arrows, and periods of exponential growth by vertical hatching.

which is the slope of the ln W curve. From this it follows that the average value of R over any given interval, (t_1, t_2) is

$$\frac{1}{t_2 - t_1} \int_{t_1}^{t_2} \frac{d \ln W}{dt} = \frac{\ln W_2 - \ln W_1}{t_2 - t_1}$$

or

$$= \frac{\ln (W_2/W_1)}{t_2 - t_1},$$

and is independent of changes in relative growth rate during the interval, no assumption being necessary as to the change of W with time.

If it is assumed that W increases exponentially with time, then

$$W_2 = W_1 \, e^{R(t_2 - t_1)}$$

11

and
$$\ln W_2 = \ln W_1 + R(t_2 - t_1).$$

From this
$$R = \frac{\ln W_2 - \ln W_1}{t_1 - t_1}$$

as before.

This second derivation is the one usually given, but it lays too much stress on the assumption of exponentiality. The first derivation requires no such assumption.

When W is plotted against time on semilog paper, changes in R can readily be deduced since

$$R = \frac{\mathrm{d}\ln W}{\mathrm{d}t} = \frac{\mathrm{d}(\log_e 10 \cdot \log_{10} W)}{\mathrm{d}t}$$

$$= \log_e 10 \cdot \frac{\mathrm{d}(\log_{10} W)}{\mathrm{d}t}$$

$$= 2.3026 \times \text{slope of graph on semilog paper.}$$

Any period over which the slope is constant is one of exponential growth. A habit of scanning log data in this way is clearly to be encouraged, for I know of no better way of arriving at a time course for R. This is because any necessary smoothing of the data is best done with their logarithms.

Incidentally, I much prefer the symbol R to the initials RGR in fairly common use. The latter is clumsy and does not lend itself to the use of subscripts.

Returning to Fig. 2.1, it will be apparent that R_W records the time course of the slope or b value of $\log_e W$. It draws immediate attention to changes in rate, to the existence and timing of maxima and minima, and it tells us whether growth is ever strictly exponential, and for how long. The only common feature of the three curves is that, over the last ten days or so, they all decrease with time at a diminishing rate towards a zero bound. This simply records the maturation processes to be expected in determinate structures. If the data continued into senescence, this would yield small negative values of R_W.

It will be apparent that the concept of relative growth rate is a valuable and sensitive aid in the interpretation of growth phenomena. Like all sensitive tools, however, it is useless, if not actually misleading, when applied to information which lacks the necessary precision.

A further indication of the usefulness of the concept is provided by Table 2.1, which lists some relative growth rates and their equivalent generation times for a wide range of organisms and for tobacco mosaic

Table 2.1. *Relative growth rates and corresponding doubling times for a range of biological systems*

System	R (day^{-1})	Generation time (days)
Bacteriophage		
Anti-*Escherichia coli* phage (37 °C)[1]	204	0.0034
Bacteria		
Bacillus stearothermophilus (60–65 °C)[2]	90.7	0.0076
Escherichia coli (40 °C)[2]	47.5	0.0146
Vibrio marinus (15 °C)[2]	12.3	0.056
Virus		
T.M.V. in protoplasts (28 °C)[3]	22.1	0.031
T.M.V. in leaf discs (28 °C)[3]	12.3	0.056
Yeast		
Willia anomala[4]	14.0	0.050
Fungus		
Aspergillus nidulans (37 °C)[5]		
Germ tube growth (in length)	17.3	0.040
Submerged culture (dry weight)	8.6	0.080
Algae		
Chlorella (T × 7115) (39 °C)[4]	6.2	0.112
Chlorella pyrenoidosa (25 °C)[4]	1.96	0.35
Scenedesmus costulatus (25 °C)[4]	1.08	0.64
Anabena cylindrica (23 °C)[4]	0.74	0.94
Angiosperms		
Linum usitatissimum (Fig. 3.1.16)		
Leaf 52, 1st maximum	1.96	0.35
Leaf 52, 2nd maximum	0.95	0.73
Triticum aestivum		
Germinating embryo[6]	0.86	0.80
Independent seedling[6]	0.151	4.5
Early sown (stage 1, Fig. 2.8)	0.067	10.4
Early sown (stage 2, Fig. 2.8)	0.052	13.3
Early sown (stage 3, Fig. 2.8)	0.024	28.4

[1] Ellis, E. M. and Delbruck, M. (1939)
[2] Brock, T. D. (1967)
[3] Otsuki, Y. *et al.* (1972)
[4] Whaley, W. G. (1961)
[5] Trinci, A. P. J. (1969)
[6] Williams, R. F. (1960)

virus. The equivalence derives from the relation $\ln 2 = gR$, where g is the average time between consecutive cell divisions, or for doubling. Thus $g = 0.693/R$. While this is a very simple relation, the conversion tables at the end of the appendix may be found useful.

In choosing the entries for Table 2.1, high values of R were sought, and the temperatures at which these were achieved are indicated. Other values give some idea of the range likely to be encountered within the main groups. Generation times for bacteria are better recorded in minutes, being 11, 21 and 81 minutes respectively for the three examples

13

of the Table. Mature trees would need to be recorded in years. The values for *Aspergillus* are not strictly comparable with the others, being based on length in one instance, and using a very artificial medium in the other. The fact that T.M.V. virus multiplies much faster in naked palisade protoplasts than in leaf discs at the same temperature suggests that the wall is a barrier to multiplication. However, the main point to be made is that, in general, relative growth rates are negatively correlated with structural complexity – highest where most of the energy is directed to the replication of a nucleoprotein or simple protoplast, and lowest where complex and highly differentiated structures are involved. Temperature and other environmental constraints are obviously important, but internal processes of organization almost certainly set the limits under optimal conditions for growth.

An elementary but valuable property of relative growth rates is illustrated in Figs. 2.2 and 2.3, and is concerned with the relation between linear and volume growth. Fig. 2.2 records the length, width, thickness and volume for flax leaf 52 over a period of eight days. The linear measurements were made as described in the appendix (Fig. A.6 and associated text), and volumes by serial reconstruction as described for wheat primordia (Fig. A.4 and text). The logarithms of the linear dimensions are adequately described by linear regressions on time, and these are given within Fig. 2.2; but the volume data have a slightly curvilinear relation with time (the broken line in Fig. 2.2). However, the relation is fairly well described by the linear regression,

$$\log V = -7.9851 + 0.40006t,$$

shown as a dotted line in the Figure, and the coefficient of t, 0.40, is very nearly the sum of those for L, W and T, 0.41. Indeed the parallel lines which enclose the volume data of Fig. 2.2 have a b value of 0.41 and represent the volume growth of ideal solids based on the linear-dimension regressions, and being rectangular prisms (the upper line of Fig. 2.2) as exemplified in Fig. 2.3, or ellipsoids contained within those prisms (the lower parallel line of Fig. 2.2).

Since the regressions of Fig. 2.2 are all based on logarithms to base 10, the b values must be multiplied by 2.3026 to give relative growth rates, as follows:

$$R_L = 0.515$$
$$R_W = 0.326$$
$$\underline{R_T = 0.103}$$
$$0.944 \quad \text{but } R_V = 0.921.$$

14

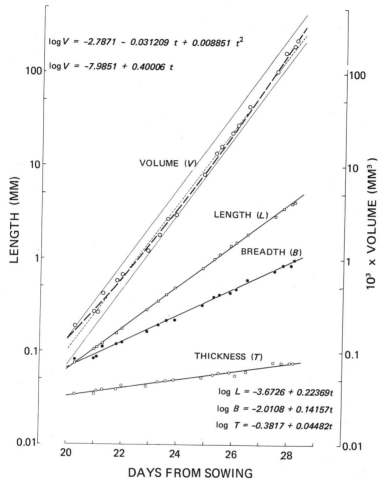

Fig. 2.2. The volume, V of the 'standard' flax leaf (see Fig. 4.1.16) and its components, length, L, breadth, B and thickness, T.

However, the curvilinear relation for volume permits estimates for days 20 and 28 at the beginning and end of the interval. They are 0.697 and 1.005 respectively, and are consistent with the minimum and maximum values for R_W shown at about these times in Fig. 2.1. However, that curve is based on the much more extensive data of Fig. 4.1.14.

The data of Figs. 2.2 and 2.3 remind us that, to a close approximation, volume relative growth rate is the sum of the three linear-dimension relative growth rates. This being so, one may ask how it is possible that

15

Linum usitatissimum

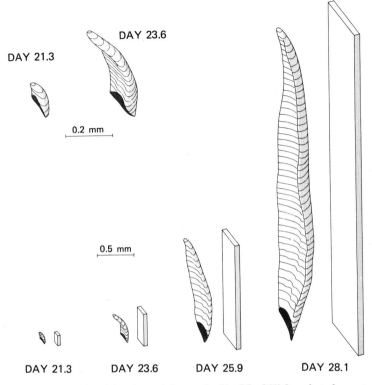

Fig. 2.3. Three-dimensional drawings of the standard leaf (leaf 52) for selected ages, together with equivalent volumes based on length, width and thickness measurements as defined in the appendix (Fig. A.6).

the volume relation can be curvilinear. The short answer is that there is a gradual change in form which is more or less independent of the linear measurements selected for the exercise. Fig. 2.3 suggests that the primordium shape changes from a triangular to a more nearly rectangular form during the period. A second implication is that R_W, on a fresh or dry weight basis, will also approximate to the sum of the three linear-dimension relative growth rates. The data also demonstrate the necessity of precision for meaningful results.

A detailed statement on growth curves and curve fitting is quite unnecessary here because the subject has been reviewed by Richards (1969) in a very thorough and scholarly manner. This document, published posthumously does not make light reading, but at each attempt I have discovered new insights and a profound wisdom which is very

16

rewarding. An example is contained in the quotation which I have put at the beginning of Chapter 8. Other useful statements on curve fitting are made in books by Steward (1968, Chapter 9) and Dormer (1972, Chapter 2), so I will content myself with a few general comments and a restatement of a practical example of my own (Williams, 1964).

There is a fascination about curve fitting to growth and population data which can all too easily become an addiction. As an eminent entomologist once said to me, 'The fit is so beautiful that the constants *must* have a fundamental meaning.' But must they? As Richards points out, mathematical functions based on simple hypotheses concerning the nature of growth are capable of reproducing the course of growth curves with tolerable accuracy, but not so accurately that a clear distinction can be drawn between one function and another. A sounder approach is to look for accuracy of fit, and to reject the notion that the mathematical form has physiological significance. The resulting curves then constitute convenient summaries of the data, and can provide a basis for interpretation without directing it. Experience will also show that precise data, especially when it covers a wide range of size will not conform to any continuous function. In such cases, only false ideas about the objective handling of data would support curve fitting in preference to careful free-hand smoothing. In certain circumstances, progressive curve fitting of the type described in the appendix, and used to construct Fig. 4.1.15, may be justified.

In general, curve fitting should be indulged in only when there are clear-cut objectives, and when the practitioner is aware of the pitfalls. It is as well to remember that it is all too easy to fit curves to poor data, and that poor data usually derive from inadequate experimental procedures. Having said this, I would still commend the use of the growth function proposed by Richards (1959), for this is so flexible that it has the best chance to provide acceptable descriptions of the results.

Computer programs are now available for fitting the Richards function to experimental data (Causton, 1969; McIntyre *et al.* 1971), and those who have the necessary mathematics should study the original paper (Richards, 1959). If one merely wishes to familiarize oneself with the properties of the function, it is well worth generating a set of curves such as those of Fig. 2.4 from the simplified equations:

$$W = [1 - e^{-t}]^{1/(1-m)},$$

for which $m < 1$ and $t > 0$, and

$$W = [1 + e^{-t}]^{1/(1-m)},$$

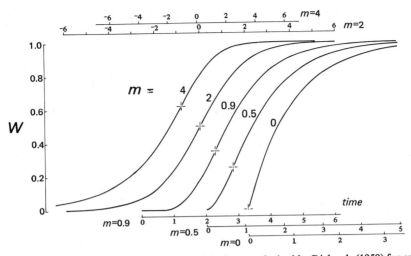

Fig. 2.4. Five examples of the infinite series of curves devised by Richards (1959) for empirical growth studies.

when $m > 1$ and t has all values. The five curves of Fig. 2.4 have a common scale of W with a limit of 1, and the time scales are adjusted so that the curves are uniformly spaced and have similar apparent maximal rates. This procedure demonstrates clearly that m is the parameter which determines the shape of the curve, for the value of W at inflexion increases with m.

A valuable property of the Richards function is that it includes three of the well known and much used growth functions as special cases. By putting $m = 0$ we get the monomolecular curve, and with $m = 2$ we have the curve of autocatalysis. Although the equations cannot be solved for $m = 1$, Richards shows that the limiting form of his general function as $m \to 1$ is in fact the Gompertz function. Thus, for values of m ranging from 0 to 1 we have curves grading from the monomolecular to the Gompertz: for values of m from 1 to 2 the curves grade from the Gompertz to that of autocatalysis; and higher values of m give curves with higher and higher points of inflexion.

Fig. 2.5 gives a set of length growth data for flax leaves to which Richards's curves have been fitted. There were from six to nine experimental values available for each curve, and in most cases the curve accounts for most of the variation. Estimates of maximum length were required before estimating the last six curves, and the rather devastating effects of a missing value at a critical time are shown by leaf 32. The experimental values of Fig. 2.5 are shown as circles if they are above the

18

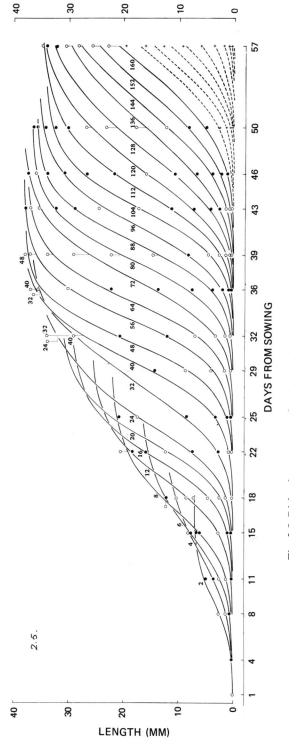

Fig. 2.5. Richards type curves fitted to the leaf length data for flax (cf. Fig. 4.1.14).

19

fitted curves, or black spots if they are below them. The device draws attention to systematic deviations from pattern. When most of the values for a given sampling occasion are of one kind there is reasonable ground for supposing the sample to be abberrantly large or small as the case may be (e.g. days 18 and 46). Where one kind of value is found systematically across a number of samples there is reason to question the appropriateness of the function in that area. Because of many difficulties of this sort, I have preferred the more subjective approach of Fig. 4.1.14 to portray the length–time relation for flax leaves. This approach is euphemistically styled 'harmonizing the data', but it has genuine interpretative value when used with due caution on extensive data of this kind.

Further practical points relating to curve fitting will emerge from a restatement of the practical example already mentioned (Williams, 1964).

The data covered by Figs 2.6–2.11 derive from an experiment designed to achieve precision for a growth study under field conditions. Only shoot growth was measured, and the environment was one of steadily increasing temperatures and lengthening days. Each point in Figs. 2.6–2.9 represents 16 sampling areas carrying, on average, 16 plants each.

In the light of the foregoing discussion the simplest way to present the data would be to link successive values with straight lines, just as in Fig. 2.10 for leaf, stem and inflorescence growth. This solution has much to commend it, though it is scarcely likely to contribute directly to understanding. The same could be said about drawing free-hand curves to smooth out the week-by-week fluctuations due to environment and sampling error.

Fig. 2.6 illustrates the descriptive power of the curves of autocatalysis, the linear form of which permits the rejection of values which cannot reasonably be accommodated. In this instance the last two points for the early-sown, and the last three for the late-sown crop were rejected, and, as will be shown later, the curves also do some violence to the early experimental values for each crop.

The fitting of third degree polynomial curves to the logarithms of the original data yielded surprisingly good results, as is seen in Fig. 2.7. The curves are plotted on an arithmetic scale to facilitate comparison with Fig. 2.6, and they give a superficially satisfying description without the need to discard any terminal values. They also do more justice to the early experimental values for each crop. For all that, close inspection shows that the descriptions are not accurate in detail, especially for the

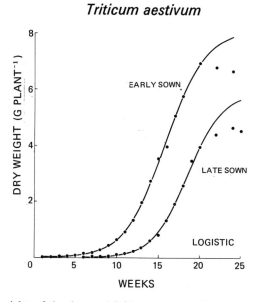

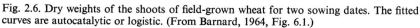

Fig. 2.6. Dry weights of the shoots of field-grown wheat for two sowing dates. The fitted curves are autocatalytic or logistic. (From Barnard, 1964, Fig. 6.1.)

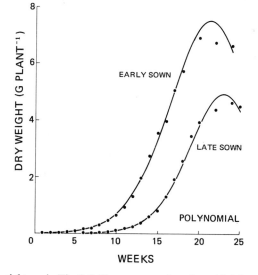

Fig. 2.7. Dry weights as in Fig. 2.6. The curves are based on third degree polynomials fitted to the logarithms of the data. (From Barnard, 1964, Fig. 6.2.)

21

Triticum aestivum

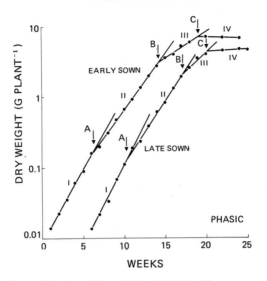

Fig. 2.8. Logarithms of the dry weights of Fig. 2.6 fitted with sequences of linear regressions. A, B and C mark the successive intersections of the linear regressions. (From Barnard, 1964, Fig. 6.3.)

late-sown crop. There is a strong tendency for sequences of points to be found either above or below the fitted curves.

The third approach to the analysis of these growth curves departs from the notion that growth is a continuous function of time. It makes literal use of the once-popular notion of phasic development, and describes growth as sequences of exponential segments (Figs. 2.8 and 2.9). Clearly there are many pitfalls in the procedure, for subjective judgements must be made as to which experimental values belong to the successive straight-line segments of Fig. 2.8. The judgements are not difficult for the early-sown crop, though very uneven numbers of values contribute to the segments. Thus segment III is defined by only four points and segment IV by three. Segments I and II for the late-sown crop are satisfactory, but segments III and IV are open to question. The last five or six values do not readily fall on straight lines at all, and the compromise solution is based all too obviously on a desire for conformity with the pattern set by the early-sown crop.

When reduced to the arithmetic scale of Fig. 2.9, the relations appear as series of exponential curves, and the descriptive power of the procedure is clearly superior to those of Figs. 2.6 and 2.7. However, its

22

Triticum aestivum

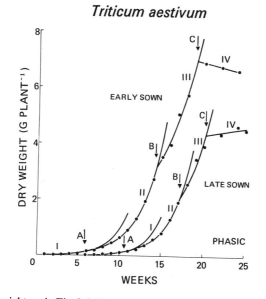

Fig. 2.9. Dry weights as in Fig. 2.6. The exponential curves are based on the linear regressions of Fig. 2.8. (From Barnard, 1964, Fig. 6.4.)

success depends to some extent on the fact that it uses up more degrees of freedom in the process. The important question is whether the form of analysis can be shown to have biological significance, and this gains support in Fig. 2.10. This presents curves for leaves (blades only), stems plus leaf sheaths, and for the inflorescences in both the early- and late-sown crops. The points A, B and C, marking the change from one exponential curve to the next, are given in all three figures, and it will be noted that B and C mark the rather sudden cessation of the growth of the leaf and stem fractions respectively. Point A is harder to identify, though independent evidence suggests that it could be the point at which dry matter is diverted to the secondary root system.

While the stage of growth approach shows signs of having biological relevance, it will already be clear that this exercise in curve fitting does not permit of an objective choice between the methods. Nor would we be much better off with the Richards function. This could be made to include all the experimental values, but would still give an unsatisfactory fit to the early values. Fig. 2.11 summarizes the data in terms of relative growth rates. The experimental values, which are for two-week intervals, suggest only a general fall with time. Superimposed on the values are the rates appropriate to the three descriptive analyses of the growth curves.

Triticum aestivum

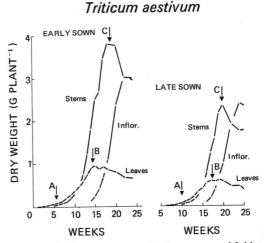

Fig. 2.10. Dry weights of the leaves, stems and inflorescences of field-grown wheat for two sowing dates. Points A, B and C are as in Figs. 2.8 and 2.9. (From Barnard, 1964, Fig. 6.5.)

Triticum aestivum

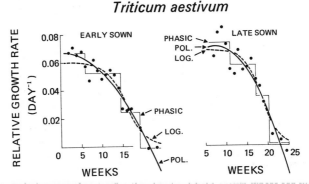

Fig. 2.11. Relative growth rates for the shoots of field grown wheat for two sowing dates. The points are experimental values for two-week intervals. The curves are appropriate to the logistic curves of Fig. 2.6, the polynomials of Fig. 2.7 and the exponential or phasic treatment of Fig. 2.9. (From Barnard, 1964, Fig. 6.6.)

Of the two pairs of continuous curves in Fig. 2.11, those based on the third-degree polynomial are again the more satisfying. Those for the logistic curves fail both at the beginnings and endings of the growth periods. On the other hand, one can scarcely claim that the experimental values uphold the stepwise fall in relative growth rate which follows from a sequence of discrete exponential segments.

Mention has already been made of recent studies of the growth kinetics of 'Marquis' wheat (Macdowell, 1972, 1973). These were

24

limited to the early vegetative phase – usually for the interval, day 7–28 – and assumed exponential growth over that period. The assumption seems justified on the evidence presented, but in a more recent paper (Macdowell, 1973), which re-examines earlier work by Friend *et al.* (1962), many of the treatments are said to have produced exponential growth for only a week. Indeed, Friend found R to decline continuously after germination. This suggests that the identification of an exponential phase from the data itself, especially on two or three points alone, is too subjective to be acceptable in an otherwise promising form of analysis. It would be safer to use biological criteria for defining the interval – from the day of emergence of leaf 2 (when seed reserve effects are no longer important) to that on which leaf 5 begins to emerge, for instance. By putting all the effort into the precise determination of W_1 and W_2 on such a basis, one would get mean values of R of which Blackman would have approved, and without an unnecessary and perhaps unwarranted assumption as to exponentiality. The point can be illustrated by a rather beautiful set of data for cocksfoot (Broué *et al.* 1967).

Data for four of seven populations studied by these authors are presented in Figs. 2.12 and 2.13. Initial and final harvests were taken when the third and ninth leaves of the primary shoots had emerged, and this was done independently for each harvest class. Because of temperature effects the inter-harvest period ranged from 21 to 82 days. The populations showed a general similarity in their patterns of response, but significant differences were also recorded. It is to be hoped that the curious interaction between day length and temperature revealed by this study will be examined further. The growth analysis into net assimilation rate and leaf area ratio (Fig. 2.13) is a first step in that direction, and it shows that population 3 behaved very differently to the others. The desirability of using biological criteria for sampling gains strong support from this experiment.

Another very promising approach to the study of plant response is exemplified by Gates *et al.* (1971, 1973). The first paper is on the effects of droughting and chilling on *Stylosanthes humilis*, and the second studies the interaction of cold stress, age and phosphorus nutrition on *Lotononis bainesii*, a component of subtropical pastures. These studies stress the need for an experimental approach that observes developmental response over a wide range of interaction, rather than by varying single factors at a time in an otherwise constant situation. This is seen to be especially necessary if data obtained with controlled environments are to have relevance to the variable milieu of the field. The authors use

Dactylis glomerata

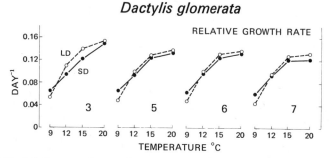

Fig. 2.12. Relative growth rates for four lines 3, 5, 6 and 7 of cocksfoot as affected by temperature and length of day (derived from Broué *et al.* 1967). In all cases, night temperatures were 5 °C lower than the day temperatures shown: LD, long day; SD, short day.

Dactylis glomerata

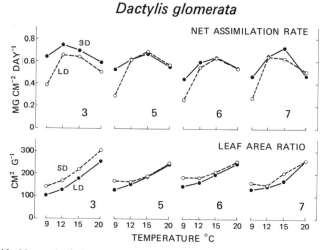

Fig. 2.13. Net assimilation rates and leaf area ratios for four lines of cocksfoot (see Fig. 2.12). (After Broué *et al.* 1967.)

both univariate and multivariate analyses when dealing with their data, and they state that multivariate analyses provide qualitative, simply expressed summaries of overall patterns at relatively low cost.

3 Phyllotaxis

'I, for my part, see no subtle mystery in the matter, other than what lies in the steady production of similar growing parts, similarly situated, at similar successive intervals of time.'

D'ARCY WENTWORTH THOMPSON (1942)

This salutary statement from D'Arcy Thompson's chapter on phyllotaxis comes immediately after an assertion that Church (1904) saw in phyllotaxis an organic mystery, a something for which we are unable to suggest any precise cause. It could also be reassuring for those who have experienced difficulty in following the erudite mathematical analysis of the subject by Richards (1951). It is amusing to note that, in another place, Richards (1948) says that Church's theoretical treatment of the subject stood little chance of sympathetic hearing, or even understanding, in a day when the average botanist prided himself on his incompetence in even the most elementary mathematics.

While it is to be hoped that botanists no longer pride themselves on such ignorance, it is a moot point whether many of us are well qualified even today, and it remains a fact that the study of the phenomena of phyllotaxis is much neglected. Those interested in the history of the subject should consult the papers already mentioned, and others by Snow and Snow (1931), Wardlaw (1949), R. Snow (1955) and Richards and Schwabe (1969). In his chapter entitled 'The Succession of Parts', Dormer (1972) also provides a valuable general commentary on the subject.

The immediate purpose of this chapter is to set out Richards's objective procedures for the quantitative description of shoot apical systems, and to use them as a basis for geometric modelling of such systems. Descriptions, including phyllotactic analyses of contrasting types of systems will occupy the whole of Chapter 4, and I will return to mechanisms when considering the possible role of physical constraint as a determinant of growth and development.

3.1 SPIRAL AND OTHER SYSTEMS OF PHYLLOTAXIS

Phyllotaxis may be defined as leaf arrangement or positioning, and its assessment is essentially a problem in geometry. So defined, it is independent of the shapes of the primordia, and even of the shape of the apex itself when all dimensions are projected onto a transverse plane. However, while making this point concerning the shape of the primordium, one should keep in mind that the size of the new primordium relative to that of the apex is of considerable significance.

Three-dimensional drawings of the shoot apex of flax, *Linum usitatissimum* are displayed in Figs. 4.1.1 and 4.1.2, the apex proper being shown surrounded by leaf primordia and young leaves up to a length of about 1.5 mm. Five of the seven apices are redrawn on a larger scale in Fig. 4.1.3 with the older leaves removed; and that for day 15 is shown here (Fig. 3.1) reduced to a 'paraboloidal' surface. Such a structure is often styled the apical cone, and this term is retained because Richards bases some of his transformations on the elementary properties of the cone. Our apical cone shows attachment areas for all primordia on the near side, and there are bulges near the apical dome which would have become primordia 31–33. The attachment areas are shown to be almost circular in outline, and they tend to make contact in such a way as to produce two sets of expanding spirals centred on the apical dome. Although the apical cone is itself an abstraction from the natural object, it is not sufficiently abstract to facilitate simple measurements for comparative studies. Its transverse projection achieves this, and is shown in Fig. 3.1. Such projections are implicit in all methods used to define the radial and tangential spacing of primordia. In practice they are usually built from serial sections, and practical considerations relating to their construction are presented in the appendix.

The projections of Fig. 3.1 show that the primordia, numbered in the order of their formation, are arranged in two sets of spirals. The most prominent set of spirals is the steeper one whose individual members differ by five (e.g. 12, 17, 22, 27); only slightly less prominent is the set of three spirals whose members differ by three (14, 17, 20). These two sets are the *contact parastichies*, a parastichy being a spiral linking primordia which differ by some constant in their order of development. These figure prominently in Church's treatment of the subject, where spiral systems are designated by the numbers of spirals in the two sets, $(3+5)$ in this instance. However, other sets of parastichy spirals may be drawn through the same primordia (e.g. 12, 20, 28), and these less conspicuous

Linum usitatissimum

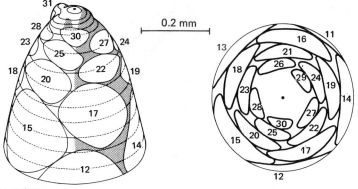

APICAL CONE   TRANSVERSE PROJECTION

Fig. 3.1. Three-dimensional and transverse projections of the apex of a 15-day flax seedling.

sets have just as much, or just as little, objective reality when it comes to the numerical definition of shoot-apical systems.

Fig. 3.2 illustrates a number of phyllotactic systems; A, B and C are extreme types of the ubiquitous Fibonacci spiral series. They are for *Brassica oleracea* $(1+2)$, *Linum usitatissimum* $(3+5)$ and *Helianthus annuus* $(34+55)$, and the series receives its name because the contact parastichies are neighbouring pairs of the Fibonacci series, in which each term (except the first two) is the sum of the two preceding terms: 1, 1, 2, 3, 5, 8, 13, 21, 34, ... Closer examination of these three diagrams reveals a weakness in description by contact parastichies, for in each case a third series of contacts is evident. For *Brassica* every third primordium is also touching; for the older primordia of *Linum* every second is touching; and for the inner part of the corona of disc florets of *Helianthus* every twenty-first is in contact. Such situations can be thought of as transition zones between spiral systems of increasing complexity. They could be designated $(1+2+3)$, $(2+3+5)$ and $(21+34+55)$ in our three examples, but the accepted terminology refers only to the two most prominent sets of intersecting spirals.

Although Fibonacci spiral systems are much more common in nature than any others, examples belonging to the so-called 'first accessory series' (1, 3, 4, 7, 11, 18, ...) are not uncommon, and some others have been reported (see below). Then, too, there are double (bijugate) and multiple (multijugate) spiral systems in which two or more primordia are initiated simultaneously. Indeed Richards saw even the distichous

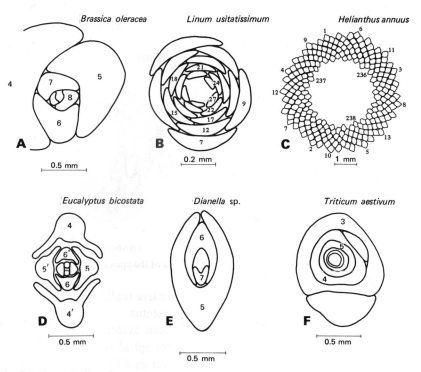

Fig. 3.2. Some phyllotaxis systems.
A. *Brassica oleracea*. Transverse projection of a low-order Fibonacci system (1+2).
B. *Linum usitatissimum*. Transverse projection of a (3+5) system. C. *Helianthus annuus*. A corona of 238 disc florets (34+55) – based on Fig. 4.9.6. D. *Eucalyptus bicostata*. Transverse section through the apex. Decussate. E. *Dianella* sp. Transverse projection. Distichous. F. *Triticum aestivum*. Transverse projection. Distichous, though not obviously so. (See Fig. 4.7.6.)

and decussate systems as limiting cases of unijugate and bijugate spiral systems respectively. Distichous systems are those in which the leaves are arranged in two opposite vertical rows, with only one leaf per node; decussate systems have leaves in pairs, each pair being at right-angles to the pair above and below. Fig. 3.2 gives examples of decussate (*Eucalyptus bicostata*) and distichous systems (*Dianella* sp. and *Triticum aestivum*). It will be noted that diagrams A, B, E and F of the Figure are transverse projections of attachment areas of leaf primordia; that C is based on the expanded capitulum of a sunflower (see Fig. 4.9.6); and D is a simple transverse section through the tip of the apex, it being difficult to develop a transverse projection in this instance (but see Fig. 4.6.10).

Some idea of the frequency of various types of spiral system may be obtained from Table 3.1, which is based on a survey of contact para-

30

Table 3.1. *Distribution of spiral phyllotactic systems within 100 families of gymnosperms and angiosperms**

Contact parastichy numbers	Angiosperm apices		Gymnosperm apices	
	Vegetative	Reproductive	Vegetative	Reproductive
Fibonacci Series				
(1+2)	45	—	1	—
(2+3)	335	35	11	7
(3+5)	53	43	22	5
(5+8)	4	25	7	2
(8+13)	1	12	2	—
(13+21)	—	11	2	—
(21+34)	—	2	—	—
(34+55)	—	1	—	—
First Accessory Series				
(1+3)	1	—	—	—
(3+4)	—	1	—	—
(4+7)	—	2	—	—
(7+10)?	—	1	—	—
Other Accessory Series				
(4+5) to (18+19)	—	25	—	—
Bijugate Series				
(2+4)	—	—	3	—
(4+6)	—	5	2	—
(6+10)	—	3	—	—
Whorled Series				
(3+3)	2	—	—	—
Totals (all axes)	441	166	50	14
Species represented	413	121	38	14

* Based on a survey by Fujita (1938).

stichy numbers conducted by Fujita (1938). The survey seems to have ignored distichous and decussate systems except where they occur as alternatives within a species. It covers approximately 500 species in 100 families of Gymnosperms and Angiosperms, but mainly of Angiosperms.

Among Angiosperms which display Fibonacci spiral systems, there is a very high incidence of (2+3) vegetative apices. Reproductive apices have a greater spread of types, with the greatest incidence at (3+5). As in the example of Fig. 3.2C, Fujita gives a value of (34+55) for a capitulum of *Helianthus annuus*, though a specimen reproduced by D'Arcy Thompson (1942) goes as high as (55+89) at its outer edge. Vegetative apices of Gymnosperms have fairly high values, about half being (3+5). What I have styled 'other accessory series' in Table 3.1 covers 25 axes in which the contact parastichies differ by unity (4+5), (5+6), ..., (18+19). In the main these are found in reproductive structures of species from the families Salicaceae and Araceae.

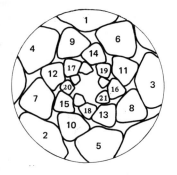

A

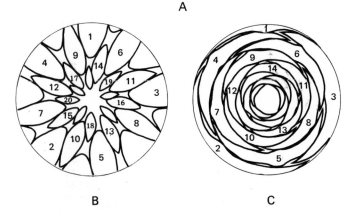

B   C

Fig. 3.3. Three distinct transverse projection patterns based on a common orthogonal 5:8 parastichy system. Divergence = Fibonacci angle; plastochrone ratio = 1.073. The contact parastichies are (5+8) in A, (8+13) in B, and (2+3) in C. (Based on Richards, 1951, Fig. 1.)

Earlier in this chapter, I drew attention to two related facts: that when phyllotaxis is defined as leaf arrangement or positioning it is independent of the shapes of the primordia; and that the inconspicuous sets of parastichy spirals have just as much, or just as little, objective reality as have the contact parastichy spirals upon which I have concentrated so far. Richards established these related facts rather neatly by setting up four ideal transverse projections, in all of which the 5-parastichies intersect the 8-parastichies orthogonally, that is at right angles (Fig. 1 *a–d* on p. 512 of Richards, 1951). Three of these projections appear here in a modified form as Figs. 3.3 and 3.4. In each, the centres of the 25 primordia are inserted at a constant divergence of 137.5° and with their distances from the centre in geometrical progression, the ratio of the progression (outwards) being 1.073. This is the required ratio to produce an orthogonal 5:8 Fibonacci system. It should be noted that the 5:8

32

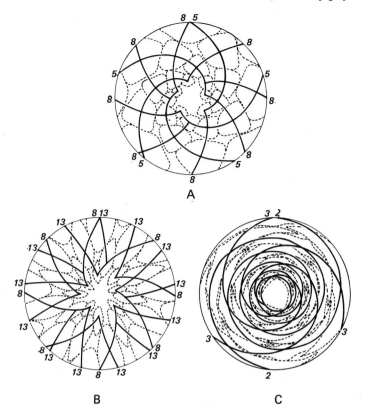

Fig. 3.4. As in Fig. 3.3 but with the appropriate parastichy lines superimposed on the transverse projection of the primordia. A, 5:8; B, 8:13; and C, 2:3. (Based on Richards, 1951, Fig. 1.)

terminology refers to intersecting sets of right- and left-handed parastichy spirals, irrespective of whether they happen to be the contact parastichies of a natural or ideal system. In keeping with the inductive approach of this chapter, I have redrawn three of Richards's diagrams with the emphasis on the shapes of the hypothetical primordia; these are only hinted at by Richards. Fig. 3.3A represents an idealized (5 + 8) orthogonal contact parastichy system such as one might expect in *Araucaria excelsa*, though those that I have dissected were not strictly orthogonal. The same system is repeated in Fig. 3.4A with the 5:8 parastichy lines superimposed on the dotted outlines of the primordia. Secondly, Fig. 3.3B represents an (8 + 13) non-orthogonal contact system, for which the 8:13 parastichy lines are drawn in in Fig. 3.4B. From my limited experience I suspect that primordia of this shape and

33

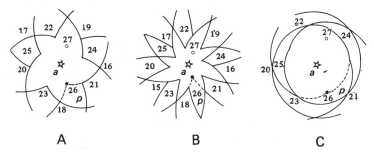

Fig. 3.5. Enlargements of the central portions of the diagrams of Fig. 3.4. A, (5+8); B, (8+13); and C, (2+3). Ideal areas for the bare apex (*a*), and for the newly initiated primordium (*p*) are indicated in each. Stars mark the centres of the systems here and in subsequent figures.

orientation are improbable, even though the transverse section through the apex of *Acacia mucronata* (Fig. 4.9.5) is superficially like it. If a transverse projection of the areas of attachment for this *Acacia* is prepared, one gets a simple (3+5) system which is very nearly orthogonal. This curious transformation of outline is brought about by the presence of short, fleshy stipules which scarcely appear in Fig. 4.9.5, but which can be seen in Fig. 4.9.4. Finally, Fig. 3.3C represents a (2+3) contact parastichy system, for which the 2:3 parastichy lines are drawn in Fig. 3.4C. Phyllotactic systems of this type are found in *Euphorbia* and in a number of Crassulacean species.

Some interesting facts concerning the intersecting curves of Fig. 3.4 can be derived from Fig. 3.7. When, as in Fig. 3.4A, the 5- and 8-parastichies intersect at 90°, the 8- and 13-parastichies of the same system (Fig. 3.4B) intersect at 45° 10', the 3- and 5-parastichies (not represented in Fig. 3.4) intersect at 135° 9', and finally the parastichies 2 and 3 of Fig. 3.4C intersect at 161° 36'. In general, it is useful to remember that, for any orthogonal parastichy system, the curves of the next higher parastichy pair will intersect at approximately 45°, and those of the next lower pair at approximately 135°.

The diagrams of Fig. 3.5 are derived from the ideal transverse projections of the previous Figure by enlarging the central portions to varying extents, so that the central areas representing the bare apex are similar in size. The diagrams will serve to introduce the time factor into the argument. The bare apices have shapes which are specific to the parastichy pairs which define them, being 5-, 8- and 3-'pointed' in A, B and C respectively. They are surrounded by similar, specific shapes which represent the maximum possible transverse areas of the primordia, and such areas are being cut off at regular intervals from the expanding bare apex. The bare areas, including those portions labelled 26 and 27, represent the apex at maximum size. Primordium 26 is then initiated and the apex is thereby reduced to its minimum size. A curious feature of the process is that, in surrendering primordium 26, the apex manages to retain precisely the same shape as before, the new figure being oriented differently – the difference being the assumed divergence of 137.5°. If, now, radial growth of the systems proceeds uniformly, and the radial distance of the inner point of 26 increases by a factor of 1.073, the presumptive inner point of 27 will have increased to unity, and the bare apex to its maximum size again. The time taken to proceed from one maximum area to the next defines the *plastochrone*, and Richards points out that the piece of apex associated with the new primordium is in the nature of a geometrical gnomon to the bare apex. A more practical definition of a plastochrone is the interval between the emergence of successive primordia.

The foregoing detailed analysis of the ideal phyllotaxis systems of Fig. 3.3 shows quite conclusively that phyllotaxis, defined as leaf arrangement or positioning, is independent of the shapes of the primordia. However, this demonstration does not destroy one's interest in these shapes, and the retention of Church's description in terms of contact parastichies is justified because it provides a bridge between the rather arid mathematical abstractions of phyllotaxis and the mysteries which still surround the genesis of form.

3.2 THE PARAMETERS OF PHYLLOTAXIS

In seeking appropriate concepts and measures for the analysis of phyllotaxis, Richards correctly insists that they should be such that their validity and usefulness do not depend on the assumptions of some particular phyllotaxis theory. Since phyllotaxis is concerned in the main with the regular repetitive pattern of leaf primordia clustered around an apex, the concepts supplied by geometry should suffice. Richards's system asserts that three parameters are necessary, namely, the *angle of the cone tangential to the apex* in the region under consideration, the *divergence angle* and the *plastochrone ratio*. The two last parameters suffice for assessment of the transverse component of the system.

The parameters of the transverse component can be defined and exemplified by reference to Fig. 3.6, which has at A the transverse projection of a seedling apex of *Brassica oleracea*. This apex has a very

Brassica oleracea

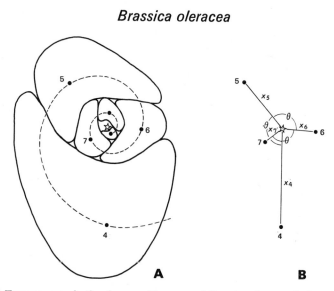

Fig. 3.6. Transverse projection for a seedling apex of *Brassica oleracea*. A shows the centres of primordia 4–9 with the genetic spiral drawn through them. B. The centres of the four older primordia, with their radii, x_4–x_7 and the divergence angle θ.

flat apical cone and is readily developed from transverse sections (see Figs. 4.3.3 and 8.1). The 'centres' of primordia 4–9 are based on their vascular bundles or their equivalent positions in very small primordia. The spiral joining the successively formed primordia is the *genetic spiral*. In Fig. 3.6B, the centres of the four older primordia are shown joined to the centre of the system. This centre is obtained by trial and error by the method given in the appendix (Fig. A.8). Fig. 3.6B supplies all the requirements for the unique mathematical description of the system. These are the *divergence angle*, θ, which approximates to 137.5° in all Fibonacci systems, and the *plastochrone ratio*, r, three estimates of which are given by the ratios for successive radii, x_4/x_5, x_5/x_6 and x_6/x_7. Ideally r is required for the zone of primordium initiation, and a procedure which aims to estimate this is presented in the appendix. Fig. 3.6 also supplies a dramatic demonstration of what is lost during the process of mathematical abstraction. Incidentally, the fact that the divergence angle is a derivative of phyllotaxis theory does not detract from its value as a parameter for describing spiral systems.

It will be evident that Fibonacci systems differ only in their plastochrone ratios, and these fall towards unity as the parastichy systems, most nearly orthogonal, rise through the successive pairs of Fibonacci

Table 3.2. *The derivation of phyllotaxis indices from plastochrone ratios for orthogonal Fibonacci systems**

Orthogonal system	Plastochrone ratio (r)	$\log_{10} \log_{10} r$	Phyllotaxis index (P.I.)
1:2	3.797	−0.2370	0.946
2:3	1.610	−0.6846	2.017
3:5	1.205	−1.0925	2.993
5:8	1.0730	−1.5145	4.002
8:13	1.0274	−1.9310	4.999
13:21	1.01035	−2.3495	6.000
21:34	1.00394	−2.7672	7.000
34:55	1.001504	−3.1853	8.000
55:89	1.000574	−3.6033	9.000
89:144	1.000219	−4.0213	10.000

* Table 1 of Richards (1951, p. 520).

numbers which characterize them. However, the ratios themselves can be rendered more meaningful by their transformation into the parallel concept of *phyllotaxis index*, P.I.

Richards noticed that the double logarithms of the plastochrone ratios for any two successive and high orthogonal Fibonacci systems differed by 0.41798, that is by $\log_{10}[(3 + \sqrt{5})/2]$. The interested reader is referred to the original paper for the derivation of this fact, but the essentials of the transformation are set out in Table 3.2. The values for P.I. in column 4 derive from:

$$\text{P.I.} = 0.379 - 2.3925 \log_{10} \log_{10} r,$$

where 2.3925 is the reciprocal of 0.41798.

These derivatives approximate closely to the successive integers, and are more useful for descriptive and classification purposes than the original plastochrone ratios. When one first encounters this transformation it seems quite unnecessarily complex, but it is no more significant in itself than, say, the need to use a logarithmic transformation before submitting growth data to statistical analysis. A number so determined for an actual apex may be called the phyllotaxis index of that apex. The index is a continuous function, but if measurements give a P.I. of 4, it may be inferred at once that the fourth Fibonacci parastichy system, 5:8 is exactly orthogonal.

Although Richards suggests that the index needs to be calculated only to one place of decimals, and that this degree of precision is sometimes unobtainable, I suggest that the above formula be used and the result recorded to two decimal places. A conversion table supplied by Richards as an appendix gives ranges of the plastochrone ratio, r and $\log_{10} r$ corresponding to each one-tenth phyllotaxis unit.

The foregoing discussion of the phyllotaxis index has been concerned solely with Fibonacci spiral systems, all of which have a divergence angle of approximately 137.5°. The manner in which this is arrived at from the opposite condition of dicotyledons will be examined in the final section of this chapter, and exemplified for flax and tobacco in Chapter 4. It is worth noting that the accuracy with which the angle is achieved increases quite dramatically with increasing phyllotaxis index.

In the case of the first accessory series, the use of the P.I. formula gives indices of 2.71, 3.65, 4.68 and 5.67 for the 3:4, 4:7, 7:11 and 11:18 orthogonal systems respectively. That is, the values at orthogonal intersection approximate to indices whose fractional part is 0.67; otherwise they are to be interpreted as in the Fibonacci systems. The divergence angle for the first accessory series is approximately 99.5°.

In bijugate systems two primordia are initiated in each 'plastochrone', so that the plastochrone ratio based on pairs may be supposed to alternate with ratios of 1.0. The phyllotaxis index for a bijugate system may be derived from that for the corresponding unijugate system by adding 1.440. This gives indices of 2.386, 3.457 and 4.433 for 2:4, 4:6 and 6:10 orthogonal systems respectively. The divergence angle for all bijugate systems approximates to 68.75° – just half the Fibonacci angle, and two genetic spirals may be drawn through the leaves.

According to Richards, systems having alternate, decussate, and alternating whorls of primordia are essentially similar, having one, two, three or more members at each node. In the alternate system two parastichies wind around the apex in opposite directions, each passing through every primordium. The system can thus be designated 1:1. The decussate system becomes a bijugate version of the alternate system, and whorled systems are multijugate versions of the same system. Phyllotaxis indices for successive orthogonal members of the series, commencing with 1:1 are: 0.057, 1.497, 2.340, 2.938, 3.402 and 3.780 (for 6:6).

So far I have only considered the transverse component of the system, but Richards's analysis also requires a sufficient definition of its third dimension – some measurement which will provide an abstract link between the two- and three-dimensional projections of Fig. 3.1. We need to be able to define the parastichy curve system as it exists on the actual surface of the apex, and Richards achieved this with his *equivalent phyllotaxis index*, E.P.I. The required measurement is the angle of the cone tangential to the apex in the region under consideration. The E.P.I. is given by

$$\text{E.P.I.} = \text{P.I.} + 2.3925 \log_{10} \sin \theta,$$

where θ is the inclination of the region to the axis of the shoot. Since $\log_{10} \sin \theta$ is negative except when $\theta = 90°$, E.P.I. is less than or equal to P.I.

A simple illustration of the application and potential usefulness of this correction to the phyllotaxis index is afforded by calculating the E.P.I. for the attachment area of primordium 30 in Fig. 3.1. This primordium is close to the region of primordium initiation, and the phyllotaxis index appropriate to the region is 3.99. This means that, on the edge of the transverse projection of the bare apex of Fig. 3.1, the 5:8 parastichies intersect almost exactly at 90°. This may not be obvious at first because the 8-parastichy is not a contact parastichy in this system. Now the angle θ for primordium 30 is 44° 24', and $2.3925 \log_{10} \sin 44° 24'$ is -0.37; so that its E.P.I. is 3.62. For any given phyllotaxis index (or equivalent phyllotaxis index), the intersection angles for relevant parastichy pairs may be read from Fig. 3.7 which was developed by Richards. This tells us that in this part of the surface of the apical cone, the 5:8 parastichies intersect at less than 90° (about 70°) and the 3:5 parastichies intersect at about 120°.

In Part II of his paper, Richards makes an analysis of the ratio of the transverse area of the bare apex to the maximum possible transverse primordial area at initiation, which is his definition of the *area ratio*. I will use the more explicit, though clumsy *apex-primordium area ratio* for this attribute of shoot apices. Reference to Fig. 3.5 will help to visualize it, for the diagrams meet the requirements of the definition. While there are practical difficulties in the way of its use, especially in determining the maximum possible transverse primordial area, it has considerable theoretical interest as a further link between analysis and form. Since the maximum and minimum figures of the bare apex (Fig. 3.5) are geometrically similar, these areas are proportional to the squares of equivalent dimensions. If at the time of initiation a and p are the apex and primordium areas respectively we have: $a/(a+p) = 1/r^2$, where r is the plastochrone ratio. It follows that the *apex–primordium area ratio*, a/p is given by $a/p = 1/(r^2-1)$. Since every value of r corresponds to a specific phyllotaxis index, it is evident that the area ratio is rigidly determined by the index. Richards shows that P.I. is

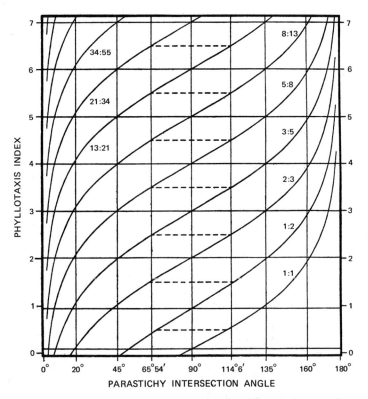

Fig. 3.7. The relation, for a divergence equal to the Fibonacci angle, between the phyllotaxis index and the intersection angles of the successive parastichy pairs. Full horizontal lines are drawn to intersect the various curves at the orthogonal values, and broken lines to meet these curves where two consecutive parastichy pairs depart equally from orthogonality. (After Richards, 1951, Fig. 2.)

directly proportional to the logarithm of the area ratio, and the relation is reproduced in modified form in Fig. 3.8.

At low indices, curves A and B diverge widely, having limits of minus infinity and zero respectively. As phyllotaxis rises, A and B rapidly approach the same straight line, C, for here the fluctuations in apical size are small, and it does not matter at what stage in the plastochrone the areas are measured. Line C has a slope of 2.3925, which is the coefficient used in transforming the plastochrone ratio into the phyllotaxis index.

At the risk of stating the obvious, I have plotted twenty experimental values for a range of seven species on Figure 3.8. Most of the values fall close to curve A, which states the relation for minimal apical area. This is because any empirical estimate is ultimately based on minimal area,

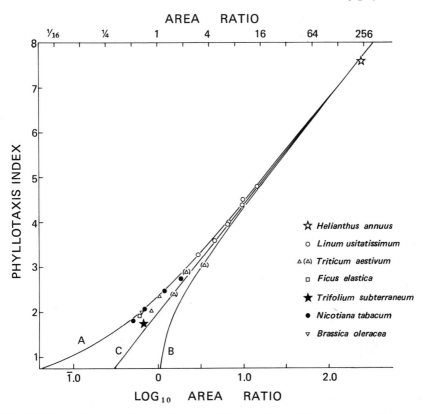

Fig. 3.8. The phyllotaxis index as a function of the logarithm of the apex–primordium area ratio. A, the relationship when minimal apical area is considered; B, that for maximal apical area; and C, that for mean apical area (based on Richards, 1951, Fig. 8). A key to the experimental values for this relation is given within the figure.

not on mean apical area as defined by Richards. Even the value for *Helianthus annuus* falls surprisingly close to the theoretical relation when one considers that it is based on measurements of a mature capitulum (see Figs. 3.2C and 4.9.6). The aberrant values for *Triticum aestivum* (triangles in parentheses) will be looked at again when examining the phyllotactic properties of this species.

Before leaving this section it will be as well to attempt a summary of its contents.

The quantitative system devised by Richards for the objective description of phyllotaxis in plants is strongly recommended. This system is set forth in a form which seeks to retain an awareness of the biological attributes of the systems thereby abstracted.

For a complete mathematical description, three parameters are

41

necessary, namely, the angle of the cone tangential to the apex in the region under consideration, the divergence angle, and the plastochrone ratio. This last is the ratio of the radial distances of two successive primordia from the central axis.

The plastochrone ratio is rendered more useful by transforming it to the related concept of the phyllotaxis index. This permits the quantitative comparison of phyllotaxis arrangements of very diverse kinds, and it is shown to be a continuously varying function.

The phyllotaxis index is rigidly related to the ratio of the transverse components of two areas, that of the central apex and that of the newly initiated primordium.

Although the statement of contact parastichies is not essential to Richards's system, their inclusion provides a very good indication of primordium shape.

3.3 THE FIBONACCI ANGLE – IRRATIONAL OR INEVITABLE?

There is little that seems mysterious about phyllotactic patterns in distichous or decussate systems, or even in whorled systems for that matter. They all have divergence angles which are natural fractions of 360°. It is the spiral systems, with their irrational divergence angles which have baffled newcomers to the subject. If Church saw in them 'an organic mystery, a something for which we are unable to suggest a cause', this did not deter him from investigating the descriptive properties of these systems with considerable skill and ingenuity. Some of his geometrical models are works of art, and they have prompted the author to try his hand at simple geometrical modelling, using Richards's system of phyllotactic analysis as a basis. However, such modelling appears to have been done long ago by Van Iterson (1907), for Richards (1948) reports that he worked out a procedure in which primordia were treated as systems of circles or spheres in mutual contact, and was able to imitate any system of phyllotaxis. Unfortunately I have not been able to sight this work, but, since Richards states that Van Iterson failed to account satisfactorily for changing phyllotaxis, I am emboldened to continue in the tradition.

[Since sending the typescript to press I have, through the kind offices of Dr J. Warren Wilson, Glasshouse Crops Research Institute, Sussex, obtained photocopies of some relevant sections of Van Iterson's book. Plate 6, Figs. 4 and 6 of that book are geometrical models of orthogonal

2:3 and 3:5 Fibonacci spiral systems, and are essentially the same as Figs. 3.12 and 3.9 respectively of this chapter. Van Iterson's methods of construction differ in detail from those given here, but it is clear that the book is deserving of closer attention than I have been able to give it. Thompson (1942), who quotes the classical literature in detail, and pays a great deal of attention to Church (1904), does not mention Van Iterson at all. Other workers, while acknowledging that the first available space theory of phyllotaxis is implicit in his constructions, leave the impression that Van Iterson's contribution was of small consequence. That this is an unjust assessment is made clear by a recent paper by Erickson (1973), which uses Van Iterson's work on the packing of spheres on cylindrical surfaces as a basis for modelling the symmetrical arrangements of protein monomers in such biological structures as viruses, flagellae and microtubules. It should not be long before the computer is used to extend Van Iterson's geometry of phyllotaxis at the shoot apex itself, and to introduce the time factor.]

Richards (1948) gives a very full account of the Fibonacci angle and its implications for spiral phyllotaxis, and shows why it is that, as the phyllotaxis index rises, the apex is led inevitably to closer and closer mean approximations to the Fibonacci angle. However, the argument presupposes the angle; it does not explain it, and Richards seems to take the angle for granted. If more than two but fewer than three primordia are present in one turn of the genetic spiral, he says, this will lead to the Fibonacci angle. If more than three but fewer than four primordia become established, this will lead to a divergence angle of approximately 99.5° – that appropriate to the first accessory system. In summing up the implications of his own model system, too, Richards (1948, Fig. 1 and p. 222) draws attention to a remarkable uniformity of pattern of angular relations, and suggests that this may go far towards explaining how the Fibonacci angle is approached with such scrupulous accuracy at the plant apex. Since Richards rejects Van Iterson's explanation in terms of space filling, or packing, it seems that he is saying that spiral systems whose members are built on the Fibonacci series of numbers *must* yield the Fibonacci angle. At the same time he is at pains to show that there is considerable latitude in the positioning of the primordia during the establishment of such systems from the decussate condition of the cotyledons. This is the point of entry for the exercise in modelling which follows later in this chapter.

In a footnote, Richards (1948, p. 223) draws attention to an early view (Wright, 1873) that the plant adopted the Fibonacci angle because this provided the best possible distribution of leaves round an axis from the point of view of illumination. Hence the Fibonacci angle became the 'ideal angle'. This point of view was criticized by D'Arcy Thompson (1942) but has been revived recently by Leigh (1972). Leigh finds that the least angular separation of leaf 1 from all the other leaves is maximum when successive leaves are separated by the ideal angle, 222.5° (360° − 137.5°). This is the requirement for minimum leaf overlap. The proof is an interesting one, even though one may not accept Leigh's assessment of the evolutionary significance of this mode of leaf arrangement.

In a chapter entitled 'The Succession of Parts', Dormer (1972) provides many insights into phyllotaxis, and reminds us that these irrational divergence angles are limiting values of the series of fractions encountered when one uses the classical, orthostichy approach to the measurement of divergence angle. Thus, 'If two leaves on the same orthostichy are separated by n internodes and if the genetic spiral makes m turns between them, the system has an angular divergence of $m/n \times 360°$'. Systems which inherently belong to the Fibonacci series yield this set of fractions: $m/n = \frac{1}{3}, \frac{2}{5}, \frac{3}{8}, \frac{5}{13}, \ldots$, which has the limiting value $\frac{1}{2}(3 - \sqrt{5}) = 0.381966\ldots$, which, in terms of the definition, implies a divergence angle of 137.50776°. . . .

Similarly for the first accessory series: $m/n = \frac{1}{4}, \frac{2}{7}, \frac{3}{11}, \frac{5}{18}, \ldots$, with the limiting value $\frac{1}{10}(5 - \sqrt{5}) = 0.276393\ldots$, and a divergence angle of 99.50155°. . . .

The second and third accessory series yield limiting values of $\frac{1}{22}(7 - \sqrt{5})$ and $\frac{1}{38}(9 - \sqrt{5})$ and divergence angles of 77.95525°. . . and 64.07936°. . . respectively.

These facts clearly support Richards by suggesting that the Fibonacci angle in particular, and the other irrational divergence angles are rigidly determined by the ideal spiral systems to which they relate. However, we still need to know how it is that initial divergence angles less than 180° but more than 120° (the limits implied above) pass rapidly and unerringly to a close approximation to the Fibonacci angle.

But to return to our modelling. The essentials are set out in Fig. 3.9, which is a model of an orthogonal 3:5 Fibonacci system. It is concerned solely with the transverse projection, centred on the star, and it assumes uniform exponential growth, along all radii. The circles represent primordia and are numbered in their order of formation.

The first thing to do in setting up the model is to establish radii at 0°, 137.5°, 275°, 412.5° (52.5°), 550° (190°), . . . using a large circular

44

Orthogonal 3:5 Fibonacci spiral system

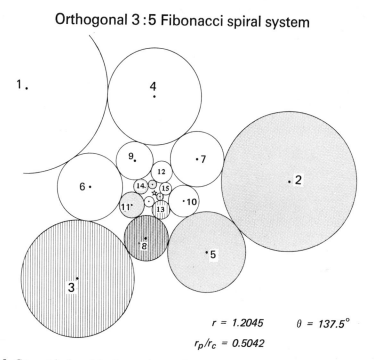

$$r = 1.2045 \qquad \theta = 137.5°$$
$$r_p/r_c = 0.5042$$

Fig. 3.9. Geometrical model of an orthogonal 3:5 Fibonacci spiral system. The members of spiral, 2, 5, 8, . . . are stippled and those of 3, 8, 13, . . . are vertically hatched.

protractor or better still, from a prepared disc with radii numbered serially in accordance with this series of angles. The remaining decimals of the Fibonacci angle, 137.50776°. . . are, of course, quite meaningless when it comes to physical drafting. Along these radii the primordium centres are stepped off in accordance with the required phastochrone ratio of 1.2045. This is simply done by entering the reciprocal of 1.2045 in the memory of a desk calculator and doing a sequence of multiplications starting with a prime radius of say fifty 0.1″ units. It pays to work on a fairly large scale on large sheets of graph paper.

The next step is to establish the radii of the primordia themselves in such a way that all the appropriate circle contacts are tangential. This is a little more difficult, if only because Richards has not already done it for us. The first thing to remember is that these radii also decrease in accordance with the plastochrone ratio. However, as will be made clear in Fig. 3.12, there is no exact solution to the problem, and a process of trial and error is appropriate. Make a reasonable estimate of the radius for primordium 1 and calculate those for 1–8 with this as base. Then

Phyllotaxis

measure the actual distances between centres on the graph, 1–4, 2–5, ..., 5–8, and 1–6, 2–7 and 3–8. Tangential contact between each of these pairs simply requires that the sum of the radii shall equal the distance between centres; so that eight estimates of a correction factor for the radii are readily obtained, e.g. $x(r_1+r_4)$ = the measured distance from centre 1 to centre 4. If the drafting has been done carefully, there will be a small but consistent difference between the values obtained for the two types of contact. However, a general mean will give a satisfactory end result, for the discrepancy is almost within a normal line thickness. All that remains to be done is to set up the corrected sequence of radii and draw the circles. Then, since the ratio of the radius of a given primordium, r_p, to its distance from the centre of the system, r_c is constant and has a value of 0.504 in this instance, we have now defined the system completely. The necessary and sufficient conditions defining the system are the divergence angle, $\theta = 137.5°$, the plastochrone ratio, $r = 1.2045$ and the additional ratio, $r_p/r_c = 0.504$. This latter constant is required by the model as such, and its precision becomes greater for systems with high phyllotaxis indices, but rapidly less for lower indices, as we shall see. The model has the virtues of simplicity and ease of construction, and will be used as a predictor for a number of basic transitions.

Attention is drawn to the inherent stability of the system, and to the perfection with which a set of circles belonging to an exponential series are thereby packed together. Then, remembering that the system is an orthogonal one, it is not difficult to see that the sets of spirals do in fact intersect at $90°$ – see, for instance, the intersection at 8 of the $2, 5, 8, 11, \ldots$ and $3,8,13,18, \ldots$ spirals in Fig. 3.9.

The very stability of the model system prompts the thought that given an approximation to its requirements in a natural system this would inevitably end up very like the model system, complete with the Fibonacci divergence angle. The necessity of this is perhaps obvious to the mathematician, but a visual demonstration may be helpful to others. Fig. 3.10 is such a demonstration, and it constitutes a simulation of the transition from the decussate condition of a pair of cotyledons to that of an orthogonal 2:3 Fibonacci system, the rules of the game can be developed as we go.

At stage 1, the cotyledons have grown sufficiently apart to accommodate the first primordium, 1, and this is sited so that $r_p/r_c = 0.73$, the figure required for an orthogonal 2:3 system. By stage 2 the diameter of 1 and the distance apart of the cotyledons have increased by a factor

46

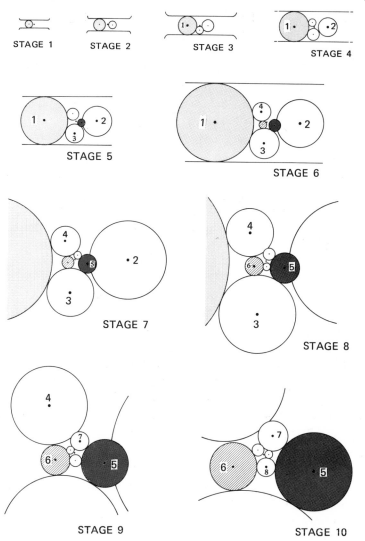

Fig. 3.10. Genesis of an orthogonal 2:3, Fibonacci spiral system from the decussate condition of the colyledons. A new primordium is generated at each stage so that the implied time intervals are successive plastochrones.

of 1.61 (the plastochrone ratio for our 2:3 system), and the second primordium has initiated on the far side of the apex and equidistant from the cotyledons. Note that growth by the factor 1.61 operates for all existing structures from one stage to the next, in what follows, so that the intervals between stages are unit plastochrones.

At stage 3 a new rule is required, and this will be that each new primordium, *n*, shall be as far removed from $n-1$ as possible, but not overlapping any older primordia. It can be tangential to the older of the two between which it arises. This places the third primordium tangential to 1, and in one of two possible positions. The choice, needless to say, will determine whether the spiral is to be clockwise or anticlockwise, and we have chosen the clockwise alternative. Stage 4 presents no problems, and places 4 between 1 and 2, but tangential to 1.

Stage 5 presents a new difficulty because the appropriate space is between 2 and 3 and is not wide enough to accommodate the new primordium. The rule can be modified so that the new primordium overlaps the older, but is tangential to the younger. This makes 5 tangential to 3, as shown. From this stage on, no new difficulties arise. At stage 6, there is just room for 6 and it is tangential to 3; at stage 7, there is almost but not quite enough room, so 7 is tangential to 5. Similarly, and in their turn, 8, 9 and 10 are fairly accurately placed between 5 and 6, 6 and 7, and 7 and 8 respectively.

Fortunately, this rather elaborate demonstration can be condensed in all essentials to the single diagram of Fig. 3.11, which only omits the cotyledons from the complete story. The final step is the measurement of the divergence angles within the model, and these are plotted to the left of the diagram. After two wild fluctuations these settle quickly to values which are quite close to the Fibonacci angle. Omitting the first two values the mean is 139° (cf. 137.5°). It should be noted that essentially the same result is achieved when the rules are varied so that new primordia are placed between or equally overlapping the primordia between which they fall. It is also recognized that the assumption of a constant plastochrone interval is too rigid, so that, in a natural system, one would expect to have early fluctuation in timing as well as in angle. In particular, the need to overlap (as in 5) would disappear if the gap were allowed to grow until the new primordium could be accommodated. It should be possible to modify the model progressively so as to explore the likely determinants of a well-defined, naturally-occurring type of transition, but our immediate purpose has been served by the demonstration that the seemingly mysterious Fibonacci angle is as inevitably a consequence of optimal packing as is the hexagonal packing of uniform spheres in an unlimited space.

Using the same set of rules and the appropriate constants for orthogonal 3:5 and 5:8 systems, I have been able to prepare similar models to that of Fig. 3.11, and to show that they, too, yield the Fibonacci angle. However, it seems more likely that the higher systems are arrived at by

Development of an orthogonal 2 : 3 Fibonacci system

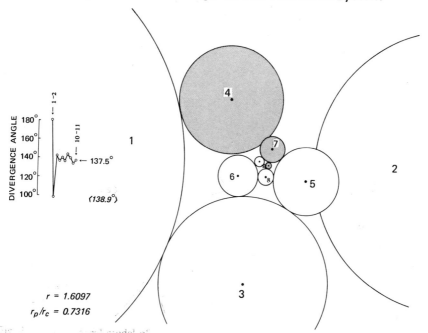

Fig. 3.11. Geometrical model of the development of an orthogonal 2:3 Fibonacci spiral system. The inset shows the progression towards the limiting divergence angle, 137.5°, and, in parentheses, the mean for the last eight values.

transition from the lower, rather than directly from the decussate condition. Tobacco and cauliflower have both been shown to start very close to an orthogonal 2:3 arrangement (Tables 4.2.1 and 4.3.1). Flax, on the other hand, has already passed the 3:5 condition by day 4, and is approaching the 8:13 condition by day 50 (Table 4.1.2).

With these facts in mind it seemed a good idea to see if the model could be used to demonstrate a progression from low to high Fibonacci systems. The result is set out in Figs. 3.13 and 3.14, but, before discussing these, it will be as well to have another look at the discrepancy found when determining an appropriate set of radii for tangential contact within these model systems. The discrepancy was small for the 3:5 system of Fig. 3.9 but is too large to be ignored in the orthogonal 2:3 system of Fig. 3.12. This was prepared in the same way, but each primordium fails to make contact with the smaller and overlaps the larger of the neighbours between which it falls. A little reflection will show that these effects arise from the curious asymmetry of each group

49

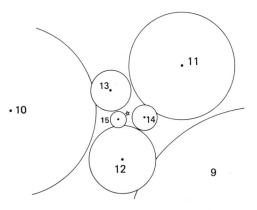

Fig. 3.12. Geometrical model of an orthogonal 2:3 Fibonacci system. This illustrates the contact discrepancies inherent in the model.

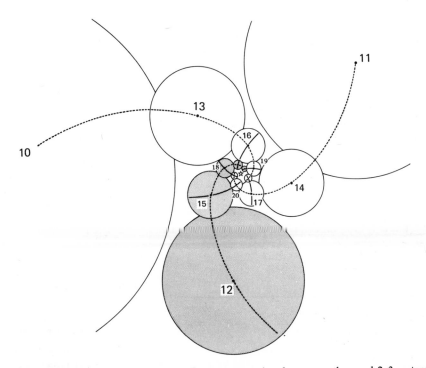

Fig. 3.13. Geometrical model illustrating the progression from an orthogonal 2:3 system to an orthogonal 3:5 system in ten steps. Three anti-clockwise contact parastichies are indicated by dotted spirals, and the beginnings of five clockwise spirals are indicated near the centre of the diagram.

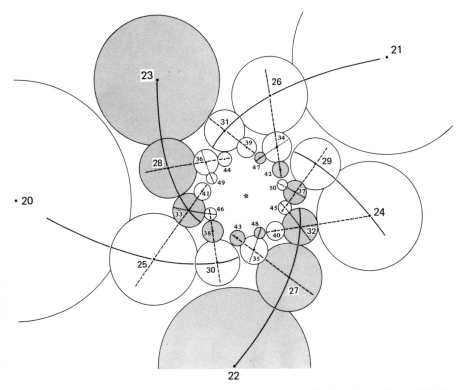

Fig. 3.14. This diagram continues the progression of Fig. 3.13 from the orthogonal 3:5 system (near primordium 20) through 5:8 and 8:13 systems to an orthogonal 13:21 system (beyond primordium 50). To assist the eye, two only of the 5 clockwise sets of primordia have been stippled. Successive sets of spirals are linked by continuous, broken, and dotted lines.

of three circular primordia relative to the centre of the system. Church recognized this problem when he used what he called *quasi*-circles in place of circles in his diagrams. They were, by definition, tangential to their four neighbours in his rigidly orthogonal systems, and were distorted circles. They do not lend themselves to modelling.

Another way of resolving the difficulty would be to depart from the Fibonacci angle sufficiently to close up the gaps. An angle of 139.2° would achieve this for Fig. 3.12, and it is noteworthy that these attempts to develop spiral systems from the decussate condition have all yielded divergence angles which are a little in excess of theoretical (see Figs. 3.11, 3.16 and 3.17). However, we decided against using such a correction in what follows because of the excessive complications of applying a diminishing correction to an already complex system.

To return to the model of Figs. 3.13 and 3.14. The objective was to simulate a progressive change from an orthogonal 2:3 system through 3:5, 5:8 and 8:13 systems, and on to an orthogonal 13:21 system. Ten steps were decided upon for each interval, and to achieve a smooth progression of values for r and r_p/r_c it was necessary to work from linear transforms of these quantities. Richards (1951, p. 563) provided ready-made steps for r in his conversion table for phyllotaxis index, and r_p/r_c is itself almost linear except from 2:3 to 3:5 and a little beyond. Graphical interpolation was resorted to in this region. The Fibonacci angle was assumed throughout the exercise and this assumption carries with it the rather unfortunate distortions considered above. Considerations of scale made it necessary to develop the model in two sections covering primordia 10–24 and 20–50 respectively and Fig. 3.14 has a scale which is approximately 22 times that of Fig. 3.13.

Fig. 3.12 can be thought of as the starting point for the model, and this shows that the 2-parastichy contacts are broken (though as explained, they should not be), but that the 3-parastichy contacts are established. The gaps between primordia which differ numerically by five are relatively large. The effect of progressive changes in the ratios of the model soon eliminates this gap however, for 13 makes contact with 18 in Fig. 3.13. The visual effect is that of the smaller members of such pairs being able to slip between their immediate neighbours to make contact with the larger. This process continues in Fig. 3.14, where primordia which differ by eight first make contact at 24–32; and those which differ by 13 at 32–45. Concurrently, of course, the number of potential sites for new primordia is increasing in accordance with the Fibonacci series; five new gaps will transform a $(5+8)$ system into a $(8+13)$ for instance. This process of breaking old and making new contact series of primordia is brought out in Figs 3.13 and 3.14 by the distinctive sets of spiral curves.

It would not be profitable to try to extract more from this rather rigid model, but sufficient has been done to indicate that it successfully 'predicts' some of the properties of naturally occurring Fibonacci spiral systems. One is reminded in particular of the progression from lower to higher phyllotaxis in flax (Fig. 4.1.8) and the implicit reverse sequence in the capitalum of the sunflower (Figs. 3.2 and 4.9.6).

Having established by simple geometrical procedures that the Fibonacci angle is an essential requirement of stable packing in Fibonacci spiral systems it is reasonable to ask if the same holds for the less common systems that were discussed earlier in this chapter. Fig. 3.15 is a model of the third member of what Richards has called the first

Orthogonal 4 : 7 first accessory system

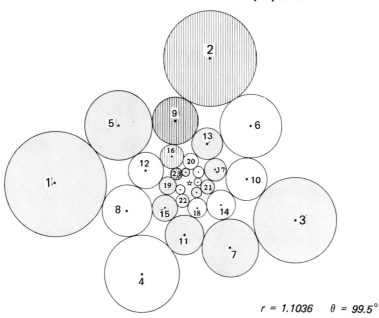

$r = 1.1036 \qquad \theta = 99.5°$

Fig. 3.15. Geometrical model of an orthogonal 4:7 first accessory spiral system. The members of spirals 1, 5, 9, ... and 3, 7, 11, ... are stippled, and those of 2, 9, 16, ... are vertically hatched. The ratio r_p/r_c is 0.375.

accessory spiral series, an orthogonal 4:7 system. Its construction followed the method given in detail for the orthogonal 3:5 Fibonacci system of Fig. 3.9, and is based upon the values of r and θ given by Richards. The ratio, r_p/r_c, was determined by trial and error as before. The model has four anticlockwise and seven clockwise spirals and, as with all members of the first accessory series, one encounters four primordia within any complete cycle of the genetic spiral. There are, of course, only three primordia per cycle in Fibonacci systems. This difference at once suggests the possibility that the development of the accessory system requires a greater degree of repulsion between young primordia, so this was kept in mind when seeking to develop an orthogonal 3:4 system from the decussate condition. The result is set out in Fig. 3.16, and it will be noticed that the cotyledons as such are omitted from the model. It is simply assumed that they have caused primordia 1 and 2 to arise on opposite sides of the apex. The placement of primordia 3 and 4, offered no problems, but they were placed as far as possible from *both* 1 and 2. Primordium 5 could only go between 1 and 3, or 2

53

Development of an orthogonal 4:6 bijugate system

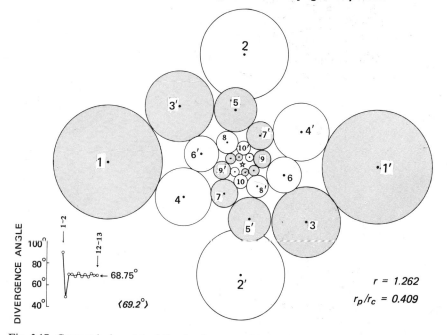

Fig. 3.17. Geometrical model of the development of an orthogonal 4:6 bijugate system. The inset shows the progression towards the limiting divergence angle, 68.75°, and, in parentheses, the mean for the last eight values.

the obvious. Enough has been said, however, to indicate that geometrical modelling has a place in the investigation of phyllotactic systems, especially for the study of transitions from one condition to another.

4 Shoot-apical systems

This chapter is devoted to the quantitative description of a number of shoot-apical systems. It does this in considerable detail, perhaps with tedious detail. Nevertheless, it supplies the bulk of the evidence upon which the thesis of the book is built, and all but the sections on wheat and subterranean clover are published for the first time. Many will be content to treat the chapter as resource material, and use it to check the claims made elsewhere. Others with a special interest in the test plants will perhaps persist, and still others will find it helpful to the planning of work on other systems.

Some attempt has been made to reduce the tedium by relegating the description of methods and procedures to the Appendix. The systems described have been studied over a period of some fifteen years so there is an unavoidable unevenness of purpose and treatment. The wheat apex was the first to be studied, followed by that for clover – a dicotyledon. Flax followed because of a growing interest in phyllotaxis for its own sake. The same may be said for *Eucalyptus*, as an example of the decussate condition, and the less extended studies of tobacco, lupin and cauliflower aimed to fill in some obvious gaps in our knowledge of spiral systems. Finally, *Ficus* was selected as an extreme example of a tightly packed apex, which could be expected to be subject to physical constraint during its long period of development.

Controlled environments are almost essential for studies of growth rate in plants, and the temperature and light regimes used in these studies were selected to give near-optimal rates of growth in each case. The experiment with *Eucalyptus* was conducted in a partially controlled greenhouse.

4.1 FLAX, *Linum usitatissimum* L.

Shoot-apical system

The drawings of Figs. 4.1.1, 4.1.2 and 4.1.3 were referred to in the previous section to illustrate the processes of abstraction needed to reduce shoot apical systems to the numerical parameters of phyllotaxis. This apex has deservedly become something of a classical object for

Linum usitatissimum

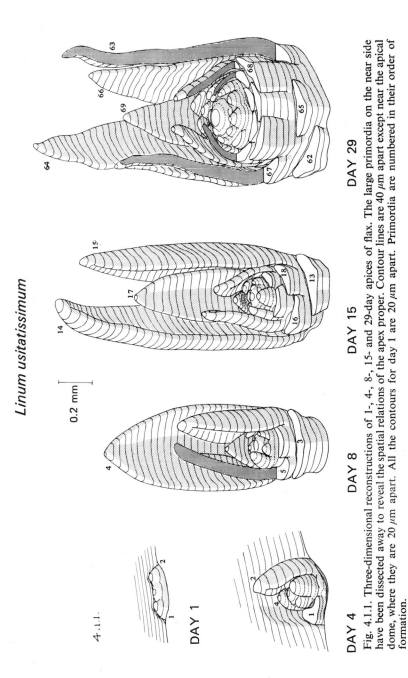

DAY 1

DAY 4　　　　DAY 8　　　　DAY 15　　　　DAY 29

Fig. 4.1.1. Three-dimensional reconstructions of 1-, 4-, 8-, 15- and 29-day apices of flax. The large primordia on the near side have been dissected away to reveal the spatial relations of the apex proper. Contour lines are 40 μm apart except near the apical dome, where they are 20 μm apart. All the contours for day 1 are 20 μm apart. Primordia are numbered in their order of formation.

Linum usitatissimum

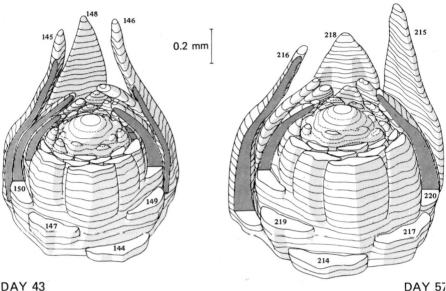

DAY 43   DAY 5i

Fig. 4.1.2. Three-dimensional reconstructions of 43- and 57-day apices of flax, similarly 'dissected'. Contour lines as in Fig. 4.1.1.

students of phyllotaxis (Esau, 1953), and illustrates many features of Fibonacci spiral arrangements. After soaking the seed for 24 hours in water, the apical dome is found to be quite small – about 0.1 mm in diameter – and flanked by the first pair of leaf primordia, which slightly overtop the apex itself. The whole structure is snugly packed into a cavity between the bases of the cotyledonary petioles (Fig. 4.1.1). Three days later the sixth primordium is present, but the whole apex is still enclosed in the cavity. It is a moot point whether this cavity grows with the apex, thereby passively accommodating it; whether the cavity is to some extent shaped by the growing apex; or whether the first pair of primordia are actually under some degree of physical constraint. From longitudinal sections it was found that the numbers of epidermal cells in a median file lining the cavity had increased from 10 to 20 between days 1 and 4 (Figs. 4.1.1 and 4.1.5A). After day 4 the positioning of new primordia quickly swings into the spiral arrangement, and the size of the bare apex gradually increases. At final harvest, when it is about to become floral, its diameter approaches 0.5 mm and there has been

Linum usitatissimum

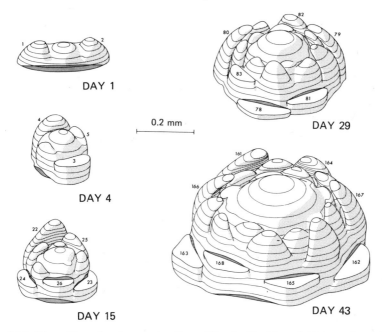

Fig. 4.1.3. Three-dimensional reconstructions of 1-, 4-, 15-, 29- and 43-day apices of flax drawn to a larger scale and with some primordia 'removed' from around the dome. Contour lines are 10 μm apart for day 1 and 20 μm for later stages.

at least a 20-fold increase in area (Fig. 4.1.2, day 57). As we shall see, this increase in size has a profound effect on phyllotaxis.

Very soon after its initiation, each leaf primordium is closely sur-rounded by other primordia, so that growth at its base is determined by the radial growth rate of the axis at that point. No such constraint appears to affect length growth, however, except perhaps that of the first pair by the cavity of the cotyledonary petioles, and a curious mutual restraint which sets in after day 29. Up to that time the primordia appear to grow quite freely into the space above the apex (Fig. 4.1.1), and there is plenty of room for marginal expansion (Fig. 4.1.4). There is however a tendency for the young primordia to curve over the apex, and by day 29 those about 0.5 mm long are so curved that they may touch at the centre (Fig. 4.1.1). By day 43 this condition extends to several layers of primordia, and is only relaxed with floral initiation (Fig. 4.1.2). While it is reasonable to suppose that the condition provides protection for a large and vulnerable apical dome, it may also constitute

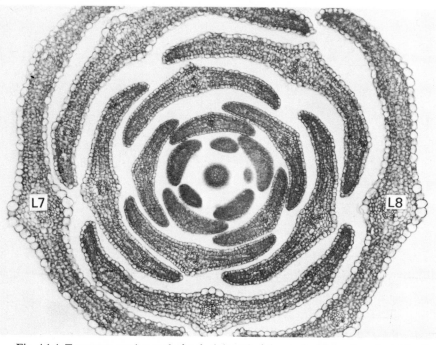

Fig. 4.1.4. Transverse section at the level of the apical dome of a 15-day flax plant. The tip of primordium 26 is to the right of the dome and the large primordium on the left is primordium 7. ($\times$ 99.)

a constraint which limits the rates of growth of younger primordia beneath these incurved primordia. A further illustration of the phenomenon is given in Fig. 4.1.6 where diagrams of transverse sections taken a little below the growing point are shown for four stages. On day 4, leaves 1 and 2 are hemmed in by the cotyledons; on day 11 the packing, though neat, is not tight anywhere; on day 25, primordia near the axis are rather crowded; and on day 39 there are three layers of tightly packed primordia around the axis. The condition on day 39 is further analysed in Fig. 4.1.19, where transverse sections above and below the apex present the story more fully. The condition is not a static one, for it is clear that as growth proceeds the outer members of the crowded primordia move away, and join the corona of larger primordia and leaves which acts as a second line of defence for the apex proper. Indeed, primordium 130 (Fig. 4.1.19) is poised between the two groups of primordia.

Yet another phenomenon calling for comment is the changing form of the leaves themselves. The earlier ones (to day 15 in Fig. 4.1.1) are

60

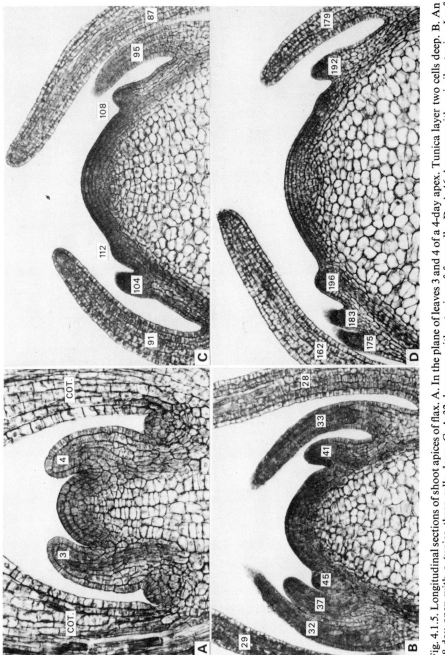

Fig. 4.1.5. Longitudinal sections of shoot apices of flax. A. In the plane of leaves 3 and 4 of a 4-day apex. Tunica layer two cells deep. B. An 18-day apex with a tunica three cells deep. C. A 32-day apex with a tunica of four cells. D. A 46-day apex with a similar tunica. Leaf numbers are estimated from other information. (All ×178.)

Linum usitatissimum

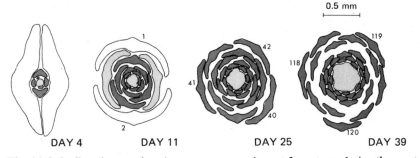

0.5 mm

DAY 4   DAY 11   DAY 25   DAY 39

Fig. 4.1.6. Outline diagrams based on transverse sections at four stages during the vegetative growth of flax. The sections are close to that part of the apical dome which is initiating new leaf primordia. The intensity of stippling is a rough guide to the intensity of meristematic activity.

Linum usitatissimum

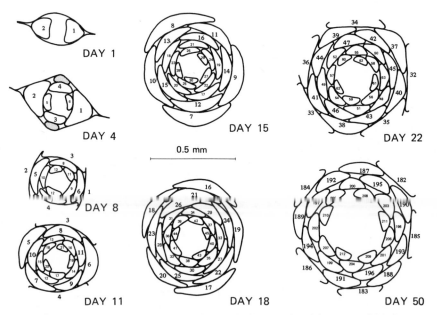

DAY 1   DAY 15   DAY 22

0.5 mm

DAY 4   DAY 8   DAY 11   DAY 18   DAY 50

Fig. 4.1.7. Transverse projections of the shoot-apical system for eight stages of development of the flax plant. The stippled areas shown for day 4 are buds in the axils of the cotyledons.

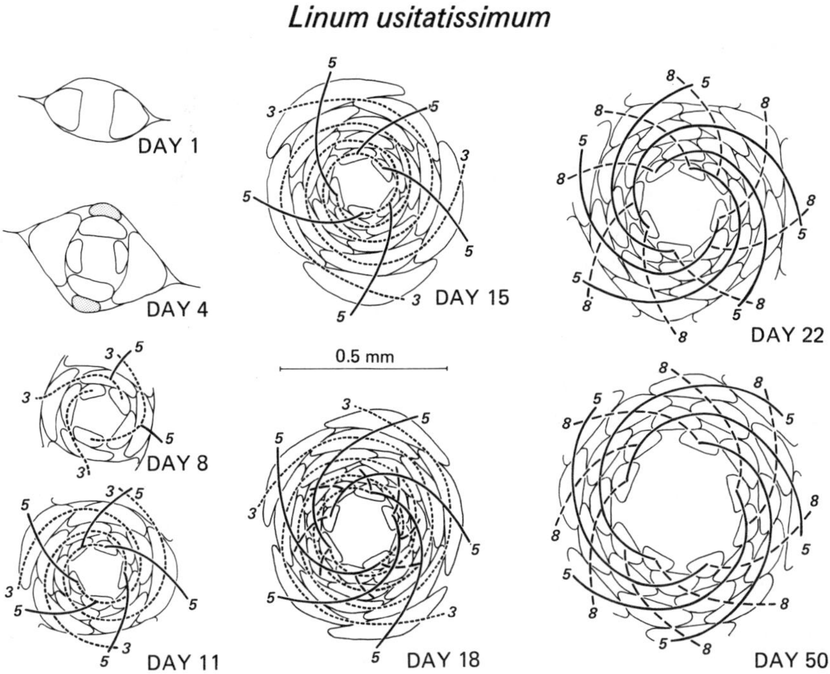

Linum usitatissimum

Fig. 4.1.8. As in Fig. 4.1.7, but with the appropriate contact parastichy spirals superimposed.

linear lanceolate and rather thick. Later they are more pointed, broader towards the base and distinctly thinner at the edges. All these properties can reasonably be referred to the changing properties of the system as a whole, and to the increasing constraints to which attention has been drawn.

Phyllotaxis

We now turn to the phyllotaxis of the system. Fig. 4.1.7 presents transverse projections of the apex for the eight stages indicated. The primordium attachment areas are numbered serially in the order of appearance of the primordia, and the youngest in each case is that for which at least one free section is present in transverse section. This means that two or three more had initiated, if the presence of a bulge on the flank of the apex is the criterion of initiation (see Fig. 3.1). However, it is difficult, if not impossible, to define the spatial limits of such early stages. The early pairs of primordia form an imperfect descussate system, with three, four or even five pairs set more or less at right angles. However, the members of the pairs quickly became subequal in size, the smaller arising slightly higher and a little later on the opposite flank of the apex. How can such a system convert so rapidly into a spiral one in which the divergence angles approximate closely to the limiting angle for the Fibonacci series? We have already seen in Fig. 3.1 that young primordia have attachment areas which are nearly circular, even though they are grossly distorted in transverse projection (Fig. 4.1.7). Furthermore, they have very similar areas at initiation and they arise on the flanks of an apical dome which is gradually increasing in size (see also Fig. 4.1.5). These are the precise conditions which underlie the explanation provided by R. Snow (1955) for the transition to spiral phyllotaxis as exemplified in the seedling of *Lupinus albus*. Snow holds that each leaf is determined in the first sufficient space that is a sufficient distance below the summit, and that one may think of its base as if it were a circular plastic disc, partially wrapped round the apical cone and free to slide down the oblique surface under gravity (cf. Fig. 3.1 above). This model may be applied with advantage to the ontogenetic sequence of Fig. 4.1.7. Between day 1 and day 4, four new primordia are formed and the dicussate system of the mature seed (day 1) tends to be maintained. Since primordia 3 and 4 are almost equal in size, and the apex must have been small when they initiated, their disc bases could scarcely do other than come to rest against the cotyledons and between 1 and 2, and on an axis at right angles to them. However, as already stated, later pairs of primordia become subequal, primordium 5 being larger than 6 in Fig. 4.1.3 (day 4). The disc base of 5 might well become fairly tightly wedged between 3 and 4 and above 2, but the disc of 6 seems to have had a choice of coming to rest against 3 or 4, leaving a gap on the other side. While such an event may not occur always at this level, it is clearly the event which determines whether the genetic spiral will be clockwise or anticlockwise thereafter.

Between days 4 and 8, seven more primordia have been added, and spiral phyllotaxis is becoming established by day 8. Nevertheless, in terms of the model, the disc bases have come to rest in strangely varied ways. Thus, 11 is at rest against 8 and 9, but does not touch 6 because the gap between 8 and 9 is too small; 12 is sitting very firmly above 7 because the gap between 9 and 10 is so large that 12 can scarcely touch them both; and 13 is firmly against 10 and above 8, but is scarcely touching 11. How, then, does the beautiful regularity of day 11, where there are 5 prominent contact parastichies running clockwise, arise (Fig. 4.1.8)? Almost without exception, the inner primordia of day 11

64

Table 4.1.1. *The plastochrone ratio in flax as affected by age*

Age (days)	Primordium interval	Ratio (r)	Age (days)	Primordium interval	Ratio (r)
1	Cot.–2	1.330*	25	63–73	1.045
4	2–4	1.189*	29	84–94	1.041
4	4–6	1.152*	32	100–110	1.044
8	4–14	1.101*	36	125–135	1.035
11	10–20	1.082	39	151–161	1.030
15	20–30	1.073	43	173–183	1.033
18	35–45	1.052	50	202–212	1.034*
22	51–61	1.049			

Minimum difference for significance ($P = 0.05$) = 0.0074.
* Not included in the analysis.   Cot., cotyledon.

are in contact with primordia $n-3$ and $n-5$, but not with $n-8$; the gap between $n-3$ and $n-5$ does not permit this.

By day 18 a further change has taken place, for the inner primordia are now in contact with all three of their neighbouring predecessors – $n-3$, $n-5$ and $n-8$. By day 22 the contact with $n-3$ is breaking, and it has quite disappeared by day 50. It is clear, too, that a continuation of this trend would have established the contact between n and $n-13$. All this is reminiscent of the discussion of diagrams A–C of Fig. 3.2, where it is shown that description of spiral phyllotaxis by contact parastichies is made difficult by the existence of transition zones where one strong and two weaker contact parastichies are present.

The changing pattern of phyllotaxis with time in flax is more clearly expressed in Fig. 4.1.8 where the contact parastichies are superimposed on the diagrams of Fig. 4.1.7. Even for days 1 and 4 it is tempting to find transient expressions of 1-, 2-, 3- and 5-contact parastichies, but there is no pattern before day 8. Even then the 5-contacts have hardly begun. Nevertheless we have a clear progression through the early members of the Fibonacci series.

Another expression of the changing pattern is in the plastochrone ratio (Table 4.1.1) which falls sharply at first and then rather slowly. Even with a replication of only two, the trend is shown to be highly significant. Table 4.1.2 gives representative data for individial axes (hence the differences in plastochrone ratio between this and the previous table). With the fall in the plastochrone ratio there is, of course, a steady rise in the phyllotaxis index. The rise is continuous and only by chance does one of them have an approximately integral value, that for day 15, which is 3.99. Such a value implies that towards the centre of this apex

Table 4.1.2. *Phyllotaxis index, apex–primordium area ratio*
and percentage cover in flax (see Fig. 4.1.7)

Age (days)	Plastochrone ratio (r)	Phyllotaxis index (P.I.)	Apex–primordium area ratio	Percentage cover*
4	1.152	3.28	2.88	76
8	1.112	3.57	4.66	84
11	1.078	3.94	6.47	72
15	1.074	3.99	6.67	77
18	1.051	4.37	9.53	83
22	1.045	4.50	9.71	69
50	1.034	4.79	14.09	73

* This is the transverse area of the primordium as a percentage of the maximum possible transverse primordial area soon after initiation. Note that the mean value over all occasions (76.2 %) is very close to that of a circle within a square (78.5 %).

we have an orthogonal 5:8 system; and this must *not* be confused with the fact that the same apex is a definite $(3 + 5)$ in terms of contact parastichies. Similarly, by day 50, the phyllotaxis index is climbing towards 5 – that for an 8:13 orthogonal system in Richards's terminology, but still only $(5 + 8)$ by Church's system of contact parastichies.

Table 4.1.2 also shows that the apex–primordium area ratio increases from a value of approximately 3 on day 4 to one of 14 on day 50. At the same time the *percentage cover*, defined as the actual area of the primordium as a percentage of the maximum possible primordial area soon after initiation (both areas in transverse projection), does not change with time. Since the transverse primordial area changes little with time (Fig. 4.1.7), it follows that, mathematically speaking, the striking ontogenetic changes in phyllotaxis are referable almost entirely to the gradual increase in the area of the bare apex. The seven values for phyllotaxis index and the apex–primordium area ratio were of course used in the construction of Fig. 3.8 above.

Turning to the other parameter of the transverse component of phyllotaxis, the divergence angle, we have in Fig. 4.1.9 a demonstration of the tendency to realize the Fibonacci angle of 137.5°. Axes of three ages are combined to make this text figure, and individual values are plotted between the successive primordium numbers to which they refer. The values for day 4 fluctuate wildly between about 180° and 90°, as is to be expected for a sub-decussate system. Those for day 11 settle down more rapidly and give a mean of 138.9° ± 1.27 for the ten values between primordia 11 and 21. The mean for the ten values between 35 and 45

Linum usitatissimum

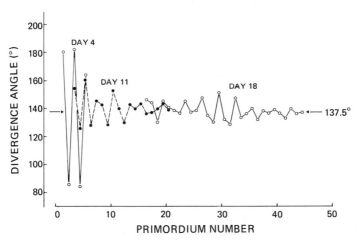

Fig. 4.1.9. Divergence angle as a function of primordium number in flax. The values for three apices of different age are here superimposed.

for day 18 is 136.8° ± 0.831, and the 50-day axis of Fig. 4.1.7 yields a mean of 137.46 ± 1.038 for the thirty values between primordia 182 and 212. The 'ideal' angle is 137.51° ..., so the agreement is excellent. However, the standard errors are surprisingly high, and can perhaps be assigned to systematic errors such as slight skewness in sectioning or errors in deriving the centre of the system. Yet another kind of deviation gains expression in the day-18 values, which have peaks at 23–4, 26–7, 29–30, 32–3 and 35–6. Reference back to the basic data for this axis shows that these peaks synchronize very well with the gaps between pseudo-whorls of three primordia; they are closer than required for regular internode spacing. This phenomenon is not uncommon in the lower parts of flax stems, but does not in my experience extend very far within a given axis.

The plastochrone, or time interval between the initiation of one primordium and the next is another attribute which is greatly affected by the gradual increase in the area of the bare apex in flax. The plastochrone is difficult to determine directly but can be derived from cumulative primordium number expressed as a function of time, as in Fig. 4.1.10. As far as day 43, the relation is very well described by the second-order polynomial within the Figure. Later the relation loses precision, presumably because of the irregular onset of floral initiation. There is no very obvious reason why a quadratic in time should success-

67

Linum usitatissimum

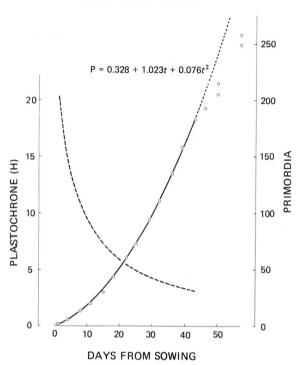

$$P = 0.328 + 1.023t + 0.076t^2$$

Fig. 4.1.10. Primordium number, P, as a function of time in flax. The values up to and including that for day 43 are well fitted by the quadratic equation, and the plastochrone curve is derived from the equation.

fully describe the relation, unless it be that space is being made available for new primordia in proportion to the square of the radius of the zone of initiation. This possibility will be examined below. The plastochrone, which is our immediate concern, is simply the reciprocal of the first differential of the P–time relation, and is shown as a broken line in Fig. 4.1.10. Specifically, the rate on day 1 is 1.18 primordia per day, which implies a plastochrone of 20 h 20 min. The plastochrone falls to 5 h on day 25, and to little more than 3 h on day 43. Gregory and Romberger (1972) have established a very similar trend in the plasto-chrone for *Picea abies* seedlings. They found values of 22 h at 15 days, but less than 6 h at 140 days. Both plants generate large numbers of small leaves on apices whose bare surfaces increase steadily during the experimental period.

The best natural markers in flax for the determination of radial

Linum usitatissimum

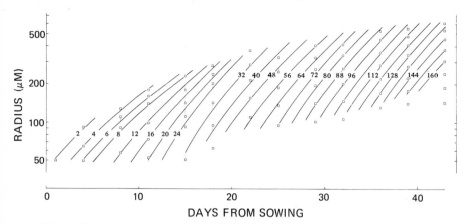

Fig. 4.1.11. Radial expansion in the vicinity of the apex of flax, based on measurements appropriate to selected leaf primordia.

growth near the apex are the creases between the primordia and the apical cone. Their distances from the centres of the system are plotted in Fig. 4.1.11 for selected leaf primordia, and the family of curves was fitted by a graphical procedure described in the appendix. The lower limits of the curves provide estimates of radial distances at initiation as previously defined. There is something like a three-fold increase in this distance – from 50 μm to 150 μm – and this is more than enough to account for the seven-fold increase in the rate of primordium production implied by the drop in the plastochrone. To that extent the data support the suggestion that space is being made available for new primordia in proportion to the square of the radius of the zone of initiation. That this should hold implies at least two other conditions: that primordium size at initiation is no very variable quantity – and this is so (Fig. 4.1.7), and that the radial growth rate within the zone of initiation is also fairly constant. That this second condition holds can be deduced from Fig. 4.1.11. The radii of this figure are plotted on a logarithmic scale, so that radial relative growth rates R_r are given by the b values or slopes of fitted curves. Estimates of initial rates so derived rise from 0.143 per day for primordium 2 to 0.209 per day for primordium 12, and the eighteen values between and including those for primordia 16 and 144 show no trend, and average 0.237 ± 0.003 per day. Even the early values are likely to have been underestimated because of the definition of initiation used.

Shoot-apical systems

A small additional point concerns the curvature of the relations of Fig. 4.1.11. It may well be that radial growth close to the zone of initiation of primordia is strictly exponential, as might be expected in such a tightly integrated system, but the data tell us that there could be a departure from this condition quite soon. This presumably implies a fairly rapid maturation of the tissues, which is scarcely surprising when one remembers that stem diameter before secondary thickening is only about 2 mm in flax, and that this is attained within a few millimetres of the apex.

The foregoing detailed account of phyllotaxis in flax has illustrated many of the properties of Fibonacci spiral systems and particularly those in which the phyllotaxis index rises progressively with time. That the various quantitative elements of the system are so closely inter-related is an inevitable consequence of the fact that Fibonacci systems are fully defined mathematically by a single parameter, the plastochrone ratio; the only other parameter of the transverse component of phyllo-taxis, the divergence angle is of course constant for all Fibonacci systems. In subsequent sections of this chapter, examples of the more common and simpler types of Fibonacci system will be examined.

Axial growth

The routine observations of this study provided not only information within the transverse plane, but also some information on extension growth within the apex. Any required axial distance can be determined from the number of 10 μm serial sections occurring between two natural markers – between half-junctions of two given primordia with the axis for instance. Such measurements provided the raw material for Fig. 4.1.12.

Attention has been drawn to the large variations which can occur in the lengths of successive internodes, so means were determined, usually of eight internodes centred on a given leaf. Thus the three values labelled 40 in Fig. 4.1.12 are based on the half-junctions of leaves 36 and 44. The more imaginative method used in Fig. 4.1.11 to portray the pattern of radial growth was rejected here in favour of the straight-line linking of points because independent evidence suggested that internode growth was a more complex phenomenon. Once again the scale is a logarithmic one, so the slopes of the lines in Fig. 4.1.12 are proportional to the length relative growth rates of the internodes ($R_{int.}$). A representative sample of these is given in Table 4.1.3.

70

Linum usitatissimum

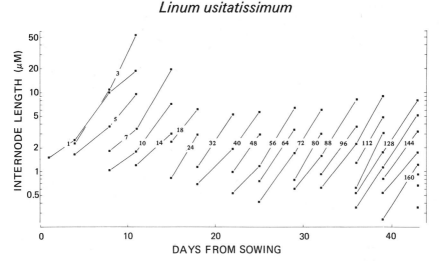

Fig. 4.1.12. Mean internode lengths near the apex of flax and appropriate to selected leaf primordia.

The first internode is a very short one and, perhaps for that reason, is the only one which gives any hint of a falling away in relative growth rate. For the remaining internodes of Fig. 4.1.12 one can derive first- and second-period values for $R_{int.}$ for only about half. However, the slopes are remarkably similar in all but the early-internode, first-period values, which have appreciably lower slopes (Fig. 4.1.12 and Table 4.1.3). Excluding the first-period values, the mean for 22 values of $R_{int.}$ for internodes 3–144 is 0.395 ± 0.015 per day. This may be accepted as a good estimate of the maximum rate of internode growth. It implies a doubling time of 1.75 days. Because all of these values *are* based on eight internodes it is not possible to get very close to the apex by this method. Indeed this is precisely why first-period values are missing from so many of the curves. An attempt to surmount this difficulty is provided in Fig. 4.1.13 and Table 4.1.4.

In principle, the procedure is the same as the classical method of demonstrating the distribution of growth in root tips, except that the markers are half-junctions of leaf primordia instead of ink spots. The markers, P_n, P_{n-3} and P_{n-5} (P13, P10 and P8 on day 8, Fig. 4.1.13) were selected because they are within the same quadrant of the apical cone of any Fibonacci system, thus minimizing errors due to any skewness of cutting. The same markers were identified at the next sampling occasion (3 or 4 days later), so that x_1 had grown to x_2, y_1 to y_2 and

Table 4.1.3. *Relative length growth rates, $R_{int.}$*
for selected internodes of flax (day^{-1})

Internode no.	First period*	Second period*
5	0.202	0.313
10	0.184	0.349
40	0.256	0.359
72	0.358	0.419
104	0.317	0.468
136	0.389	0.385

* These periods can be identified by referring to Fig. 4.1.12.

Table 4.1.4. *Axial relative growth rates,*
R_a within the apical cone (day^{-1})

Tip of dome to	Time interval (days)								
	8–11	11–15	15–18	18–22	22–5	25 9	29 32	32 6	Mean
P_n*	.292	.305	.499	.305	.334	.347	.397	.344	*.353*
P_{n-3}*	.257	.293	.366	.289	.268	.355	.358	.317	*.313*
P_{n-5}*	.216	.285	.416	.270	.304	.323	.388	.308	*.314*
Mean	*.255*	*.294*	*.427*	*.288*	*.302*	*.342*	*.381*	*.323*	*.327*

* P_n is the half-junction of the most recently initiated primordium at the beginning of each time interval. P_{n-3} and P_{n-5} are the third and fifth half-junctions below this one. See also the diagrams of Fig. 4.1.13.

z_1 to z_2 in the interval. The axial relative growth rates of Table 4.1.4 are based on these increments. Although it was not possible to obtain true replicates, we thus ended up with three related values for each of eight successive time intervals.

Although statistical treatment is quite out of order, this set of data has a number of interesting features. First, the values with time 'bounce' rather fiercely simply because they carry the full load of inter-plant variability. There is no obvious trend with time. Second, the sets of three values within interval are remarkably stable, simply because they are based on sequences of like events *within* a biological system. In particular, and with only one exception, the R_a value for the small upper part of the cone is numerically greater than the other two estimates. Is this difference real? If it is, then the internodes between P_n and P_{n-5} are growing at a still lower rate, say, 0.27 per day. But it has already been shown (Table 4.1.3) that $R_{int.}$ for similar internodes can have values of 0.20 quite close to the apex, subsequently to rise to 0.40 per day.

The outcome of this elaborate and rather speculative piece of analysis

Linum usitatissimum

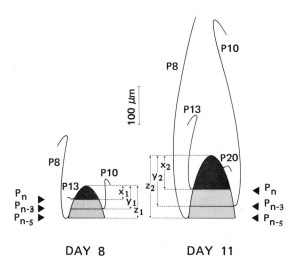

Fig. 4.1.13. The determination of relative length growth rates within the apical cone. The diagrams are based on measurements for days 8 and 11. P_n is the level of half-junction of the most recently initiated leaf for the interval (L13 in this case). See text for further explanation.

is that early internodes experience a complex sequence of relative rates of extension growth – high in the dome, falling in the sub-apical region to rather low minima, then rising to maxima which may well be higher than the rate in the dome itself, and then, of course, falling to zero as the internode matures. Fig. 4.1.12 also tells us that later internodes will be less, if at all, subject to the depression in the subapical region. It is tempting to suppose that these differences between early and late internodal growth patterns is somehow conditioned by the dramatic changes in the morphology of the subapical region (Figs. 4.1.5, 4.1.1 and 4.1.2). To describe and sort out such a sequence of events would call for a level of experimental precision considerably greater than that achieved here and, of course, proper replication.

Leaf growth

An array of length–growth curves for representative leaves is presented in Fig. 4.1.14, which includes small-scale copies of the drawings of Figs. 4.1.1 and 4.1.2. These provide reminders of the state of the apex at the selected times. The mature length is only 8 mm for the first pair of leaves, but this increases fairly rapidly to a maximum of 38 mm for leaf 48.

73

74

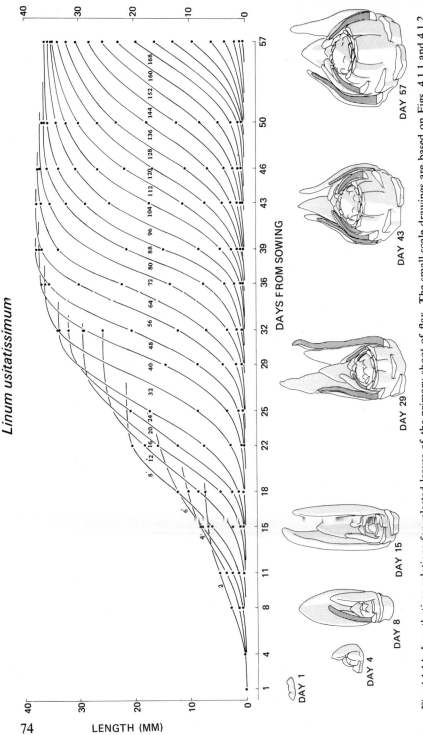

Linum usitatissimum

Fig. 4.1.14. Length–time relations for selected leaves of the primary shoot of flax. The small-scale drawings are based on Figs. 4.1.1 and 4.1.2 and indicate the state of the apex at the stated times.

Similar lengths are then maintained within the limits of the observations. At first sight the forms of the sigmoid curves of Fig. 4.1.14 do not seem to vary, but closer scrutiny suggests that this is something of an illusion. Up to and including the curve for leaf 56 the points of inflexion are high on the curves, with curvatures greater above than below that point. This condition rapidly reverses for later leaves, as the points of inflexion fall lower and lower. This change will be seen to correlate well with the progressive crowding of the leaves above the apex. The same sets of length data have been discussed in Chapter 2, where they are fitted with the flexible growth curves suggested by Richards (1959).

Figure 4.1.15 combines all the available data for volume and fresh weight of the primordia and leaves, and uses sixteen common sets of values to unite them under a common scale of estimated fresh weight. The data extend over no fewer than six logarithmic cycles so it is essential to examine them on a logarithmic scale. However, it was the changes in volume and early fresh weight which were most in need of objective handling, and it was soon realized that normal procedures of curve fitting were not applicable. Furthermore, it was considered that the simple point-to-point linking of values, such as underlies the preparation of Fig. 4.1.14, would not do justice to the data, so a fairly elaborate system of progressive curve fitting was adopted. This is explained and justified in the appendix. The procedure yielded sets of times of attainment of specified volumes, and these were linked by straight lines as in Fig. 4.1.15.

It would be tedious in the extreme to attempt a detailed interpretation of this complex figure. Perhaps it is sufficient to state that it is the basis of the relative growth rate story of Fig. 4.1.18, and to seek the reader's concurrence in recognizing that leaf growth is no stereotyped phenomenon in the flax plant. Why does leaf 2 grow differently from leaf 4? Why do the curves crowd together between days 35 and 40, but are spaced out again between days 45 and 50? Do these things mean that individual leaf growth is to some extent subject to the general constraints of the system to which they belong?

Having gone to so much trouble to determine the overall pattern of leaf growth, it may seem odd to wish to establish the time course for a 'standard' leaf of flax. However, it will be clear that the individual curves of Fig. 4.1.15 are somewhat lacking in precision. They hold each other up well enough but they do not inspire confidence when it comes to detail. Is the gentle positive curvature of most of the earlier leaves a real phenomenon for instance, or is growth then strictly exponential?

75

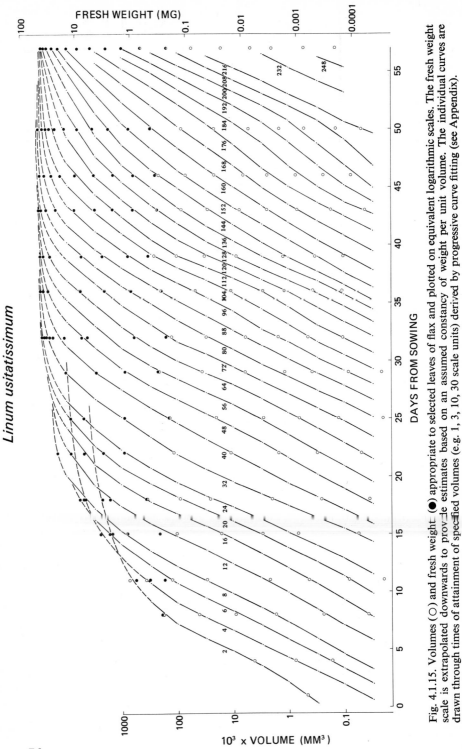

Linum usitatissimum

Fig. 4.1.15. Volumes (○) and fresh weight (●) appropriate to selected leaves of flax and plotted on equivalent logarithmic scales. The fresh weight scale is extrapolated downwards to provide estimates based on an assumed constancy of weight per unit volume. The individual curves are drawn through times of attainment of specified volumes (e.g. 1, 3, 10, 30 scale units) derived by progressive curve fitting (see Appendix).

Linum usitatissimum

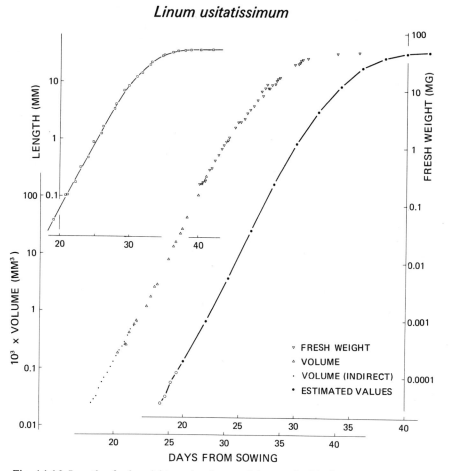

Fig. 4.1.16. Lengths, fresh weights and volumes of the standard leaf (equivalent to L52). For explanation see text.

Then, too, the experimental values are so far apart in Fig. 4.1.15 that genuine kinks could be missed altogether. Fig. 4.1.16 attempts the construction of a precise curve for leaf 52. This leaf is among the earliest to attain the maximum length, and seems unlikely to have been affected by the crowding which sets in later.

The length growth curves for leaves 40, 48, 56 and 64 happen to be almost identical (Fig. 4.1.14) and their logarithmic values are super-imposed in the upper left-hand corner of Fig. 4.1.16. This gives exponential growth from day 19 to day 28 followed by a simple curve to an upper asymptote on day 40. The curve itself provides the necessary

Linum usitatissimum

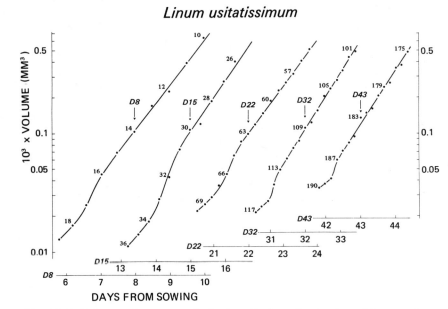

Fig. 4.1.17. Volumes of very young leaf primordia from shoot apices of the ages shown (D8, D15, D22, D32 and D43). They are plotted in such a way as to give indirect evidence of growth rates at those times.

length–time relation for plotting all available individual values for fresh weight and volume for primordia between numbers 40 and 64. The whole procedure is that of age equivalence, full details of which are given in the appendix.

Even this procedure yielded no information about growth during and immediately after leaf initiation, so this was sought by an indirect procedure which assumes that the young primordia on a limited section of an axis are equivalent to successive stages in the development of a primordium at its centre. In Fig. 4.1.17, five sets of volumes for axes harvested on days 8, 15, 22, 32 and 43 are plotted so as to simulate the volume growth of primordia 14, 30, 63, 109 and 183 respectively; these are the most recently 'initiated' primordia at the time of harvest. The time scales are derived from the plastochrone curve of Fig. 4.1.10. Two or three of the earliest volumes on each of the curves of Fig. 4.1.17 constitute extrapolations back on to the bare surface of the dome, a feat which would not have been possible without the foregoing detailed analysis of the phyllotaxis of the system.

A further set of indirect estimates of volume is plotted in Fig. 4.1.16 and is based on an axis harvested on day 18. The set links almost

78

Linum usitatissimum

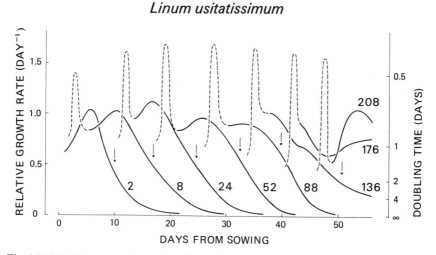

Fig. 4.1.18. Relative growth rates for selected leaves of flax (2, 8, 24, . . .). The early, rather speculative portions of the curves are dotted, and the continuous lines reflect changing slopes of the relevant growth curves of Fig. 4.1.15. The arrows mark the times at which the leaves 'emerge' from the terminal bud.

perfectly with the age-equivalence data, a fact which provides some assurance of validity. It also exhibits the same brief acceleration in growth rate as is shown in varying degrees in Fig. 4.1.17. In all cases this acceleration correlates well with the timing of the appearance of the bulge on the flank of the dome. This phenomenon might have been missed altogether if one had to rely on direct methods of investigation. Its existence implies a momentary rise in R_v to a very high level (Fig. 4.1.18).

Returning to Fig. 4.1.16, the reconstruction of the growth curve for leaf 52 is completed by calculating mid-values from linear regressions for successive overlapping, 4-day arrays of values in the manner recommended by Williams and Rijven (1965) and set out in the appendix.

The implications of a great deal of this descriptive matter relating to leaf growth at different levels on the flax stem can be brought out via the concept of relative growth rate, R_v as in Fig. 4.1.18. The curve for leaf 52 is the most reliable one, and features a rapid rise to a high peak on day 19, when initiation was proceeding. The maximum value could be as high as 1.7, with a doubling time of little more than 10 h. The rapid decline to 0.80 is followed by a slow rise to a second maximum of 0.95. Even this value implies a doubling time of less than 18 h and could be

79

Linum usitatissimum

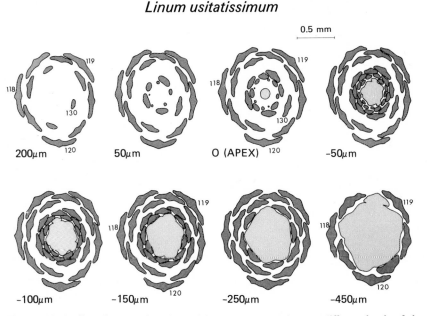

Fig. 4.1.19. Outline diagrams based on eight transverse sections at different levels of the same 39-day apex. Note the close packing of all primordia younger than primordium 130.

contributed to by the activity of marginal meristems. The second decline continues for 6 or 7 days before the emergence of the leaf from the bud, and then continues to zero in a manner to be expected in a maturing organ.

The R_v curves for leaves 2, 8, 24 and 88 are essentially the same as that for leaf 52 except that leaf 2, as already indicated, was initiated during embryogeny and shows no initial peak. Its rate on day one is also lower than for later leaves, a fact which is tentatively attributed to the constraint imposed by growth within the cavity at the base of the cotyledonary petioles (Fig. 4.1.11). The curves for leaves 136, 176 and 208 do not fit any obvious pattern; that for 136 has no secondary maximum, this maximum is much delayed in 176, and is exaggerated in 208. These vagaries in rate are believed to stem from the curious crowding of the leaf primordia which sets in soon after day 30. The nature of the crowding is well demonstrated in Fig. 4.1.19, from which it is not difficult to infer that participating leaves of different ages might behave differently.

4.2 TOBACCO, *Nicotiana tabacum* L.

Shoot-apical system

In the shoot-apex of tobacco we have another example of a Fibonacci spiral system; one which is simpler than that of flax, but which exhibits the transition from the decussate arrangement of the early seedling in a more definite and striking manner. Since the economically important part of the tobacco plant is the leaf, it is not surprising that some work has been done on its development. Avery (1933) presented an excellent series of drawings which illustrates the genesis of the blade from the thick peg-like primordium, and the subsequent production of the system of lateral veins. More recently Hannam (1968) has completed a careful quantitative study of the growth of the vegetative shoot and, using the procedure of serial reconstruction, has followed the volume changes in successive leaf primordia. The three-dimensional drawings of Figs. 4.2.1, 4.2.2 and 4.2.3 have since been developed from the collection of slides upon which the volume studies were based.

The shoot apex of tobacco is so tightly embedded in a mass of large white hairs that it is a difficult object to examine under a dissecting microscope. Fortunately, this fact did not interfere with the preparation of three-dimensional drawings, for the epidermal limits of the primordia are quite distinct in transverse section. Fig. 4.2.4 shows the hairs between the larger primordia though, because their large vacuolate cells are partially collapsed by fixation and embedding, the tightness of the packing is not as obvious as it is *in vivo*. In the immediate vicinity of the apex there are very few hairs, and those that are there are immature.

Fig. 4.2.1 provides two general views of a 25-day apex in which the eleventh leaf primordium is approaching a length of 5 mm. Various stages in the genesis of leaf form are clearly depicted by the younger primordia, and the massive leaf bases and mid-veins are much in evidence. These imply a rate of radial growth which is higher relative to length growth than is the case in flax. That the whole of the apex is packed inside and out with massive hairs implies that the relative growth rates of these hairs are greater than those of the primordia from which they arise. Even primordium 21 of Fig. 4.2.4, which is only 0.4 mm long, is producing these hairs.

Doubtless because tobacco seed is so very small, the shoot apex is organized quite slowly. A 5-day flax apex has seven or eight primordia on a sizeable apical cone, but a 5-day tobacco seedling has one primor-

Nicotiana tabacum

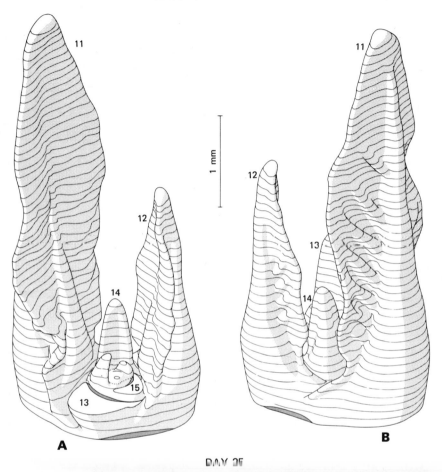

D.AY 25

Fig. 4.2.1. Three-dimensional reconstructions of a 25-day apex of tobacco. A. As seen after removal of primordia 13 and 15 to reveal the rather flat apical dome, flanked by primordia 16 and 17. B. The same from the far side to show the heavy mid-veins of primordia 11 and 12, and the genesis of lateral veins. The contours are 80 μm apart.

dium less than 0.1 mm long and a small rather flat area between the cotyledons which must be regarded as the residual apex (Fig. 4.2.2, day 5). The drawing for day 6 is a little difficult to interpret, for, in the absence of any clear distinction between an apex and a second primordium, the whole of the smaller bump must be regarded as the apex. Nevertheless the first of the two drawings for day 7, with its very small residual apex between primordia 1 and 2 suggests that virtually the

Nicotiana tabacum

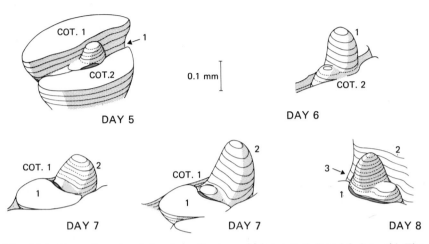

Fig. 4.2.2. Three-dimensional drawings of apices which are 5, 6, 7 and 8 days old. The continuous contours are 20 μm apart; the dotted contours reduce the interval to 10 μm where they are shown. Cot., cotyledon.

whole of the smaller bump on day 6 was about to become primordium 2. The second drawing for day 7 shows a more clearly defined apex, and the drawing for day 8 shows primordium 3 and a large apex, most of whose surface will be incorporated in primordium 4.

The timing of the ten stages of Figs. 4.2.2 and 4.2.3 should not be taken too literally. The natural variability of the material is considerable, and the drawings are simply to be regarded as representative of significant changes through which all apices are assumed to pass. Thus the three drawings for day 13, though based on seedlings of the same batch, clearly illustrate a sequence of events relative to the initiation of primordium 7. That sequence shows two important changes. First, the three primordia have already adopted divergence angles close to the Fibonacci angle, and, secondly, the apical dome is more obviously retaining its integrity from one plastocrone to the next than was the case earlier. By this I mean that primordium 7 at no time appears to engulf the whole of the apical surface, though the apex–primordium area ratio must be quite small at the time of minimal apical area (cf. Fig. 3.8).

The drawings for days 25 and 31 show quite dramatic increases in the size of the dome, particularly in that part of its surface which is not engulfed by the primordia. When these two drawings were first prepared, however, I was inclined to doubt what they seemed to be saying. How

Nicotiana tabacum

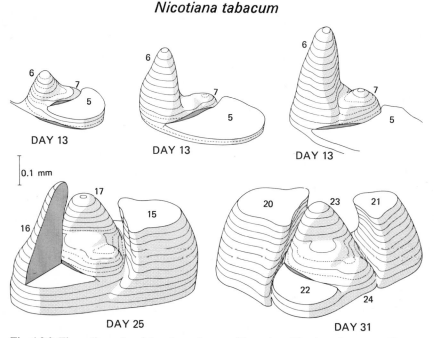

Fig. 4.2.3. Three-dimensional drawings of some older apices. The three for day 13 illustrate steps in the genesis of primordium 7. Note the considerable extent to which primordia 17 and 23 encroach upon the apex proper on days 25 and 31. Contours as in Fig. 4.2.2.

could primordium 17 on day 25, and primordium 23 on day 31 appear so high on the dome when a younger primordium was in each case at a lower level? Having ruled out the possibility that the axes were skew cut, the observation had to be accepted as fact. As we will see below, the initial growth rate of the primordium is very great and, for a short time, it must literally incorporate some of the superficial cells of the dome into its adaxial surface. In so doing the intersection of the two surfaces must move up the more slowly growing dome, only to be halted at about the time the intersection between the dome and the next primordium is generated and begins its march up the other side of the dome. Some may find this interpretation rather difficult to accept, but the alternative notion of a 'moving' apical centre has less to commend it.

Two other observations may be made here. First, that the sites of axillary buds soon to be formed (see Fig. 4.2.6) are at the centres of the lines of junctions between primordium and dome, and, secondly, that the primordia not only encroach on the dome, but rapidly fill the angles

84

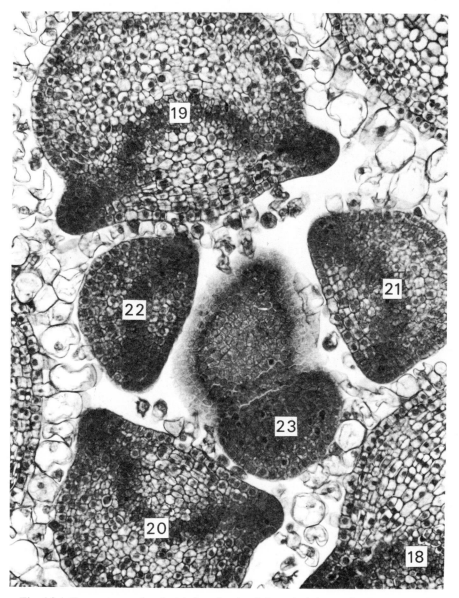

Fig. 4.2.4. Transverse section just below the tip of the apical dome of a 31-day tobacco plant. The primordia are numbered in their order of formation. Note the very large hairs between the older primordia. (× 205.)

Nicotiana tabacum

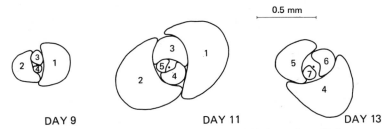

Fig. 4.2.5. Transverse projections of 9-, 11- and 13-day apices of tobacco.

Nicotiana tabacum

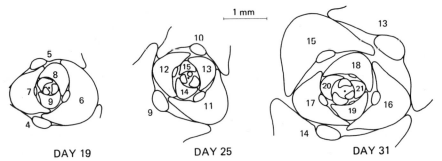

Fig. 4.2.6. Transverse projections of 19-, 25- and 31-day apices of tobacco.

between themselves and their neighbouring older primordia. It is certainly quite remarkable that, within all this competition for space, the apical dome is so well able to maintain its integrity.

Phyllotaxis

The transverse projections of Figs. 4.2.5 and 4.2.6 and the data of Table 4.2.1 exhibit many features in common with the phyllotactic properties of the flax apex. However, the transition from the decussate to the spiral arrangement of the seedling primordia is more rapid, being virtually complete with the appearance of primordium 5. The first four divergence angles for the 11-day axis are 180°, 123°, 138° and 136° (Fig. 4.2.5), and mean values for subsequent stages are sufficiently near to the ideal Fibonacci angle. The contact parastichies of Fig. 4.2.7 show that the system on day 11 is best described as (1 + 2), and that on day 31 as (2 + 3) in Church's notation. However, primordia 1 and 4 are also in

Nicotiana tabacum

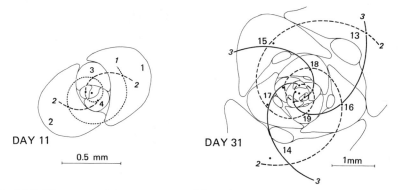

Fig. 4.2.7. The appropriate contact parastichy spirals superimposed on the transverse projections of 11- and 31-day apices.

Table 4.2.1. *Phyllotactic properties of the tobacco apex*

Age (days)	Mean divergence angle	Plastochrone ratio (r)	Phyllotaxis index (P.I.)	Apex–primordium area ratio*
9	—	1.780	1.82	0.51
11	—	1.619	2.01	0.63
13	—	1.600	2.03	0.66
19	140.1 ± 1.16	1.418	2.34	1.09
25	138.9 ± 2.15	1.362	2.47	1.17
31	139.1 ± 2.84	1.270	2.73	1.88

* Based on the 4 primordia nearest the apical dome.

contact on day 11, so the three anticlockwise spirals are on the way. Table 4.2.1 also tells us that for days 11 and 13 the phyllotaxis index is almost exactly 2, the value for an orthogonal 2:3 system in Richards's terminology. In a later chapter I will examine the possibility that this fact underlies the very rapid adoption of the Fibonacci spiral system by the tobacco apex.

Fig. 4.2.7 shows once again how very dependent the contact parastichies are upon primordium shape, and how unsatisfactory such a system is for objective description. Comparison of Table 4.2.1 with Table 4.1.2 can only confirm the value of Richards's parameters for this purpose. Both tables show that the plastochrone ratio and its dependent concepts of phyllotaxis index and apex–primordium ratio may exhibit similar time trends in different parts of the potential ranges of these parameters.

Nicotiana tabacum

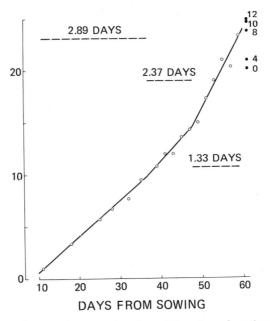

Fig. 4.2.8. Primordium number as a function of time in a short-day mutant tobacco. Plastochrones based on the linear regressions are shown as horizontal lines. Final values are for plants which received 0, 4, 8, . . . inductive cycles. Based on Hopkinson and Hannam (1969, p. 282).

The tobacco data differs quite markedly from that for flax with respect to the plastochrone. This interval falls rapidly with time in flax (Fig. 4.1.10) but only slowly in tobacco. This is clearly seen in Fig. 4.2.8, based on the data of Hopkinson and Hannam (1969) for a short-day mutant tobacco. Prior to floral initiation (day 49) the linear regressions for day intervals 8–39 and 39–49 yield plastochrones of 2.89 and 2.37 days respectively, with no significance attachable to the difference. Only after floral induction does the plastochrone fall to 1.33 days. We can deduce from the marked fall in the plastochrone ratio during vegetative growth in tobacco that the radial relative growth rate must fall with it, for the rate is proportional to the natural logarithm of the ratio when the plastochrone is constant. The radial relative growth rate is constant in flax throughout vegetative growth (see Fig. 4.1.11 and related text).

The plants to which all the drawings and also Figs. 4.2.9 and 4.2.10 of this section refer were grown with a 32°/26 °C temperature regime (Hannam, 1968), whereas those for Fig. 4.2.8 were grown with a

88

Nicotiana tabacum

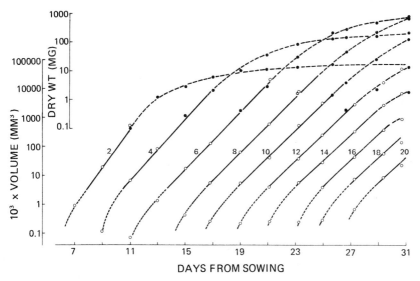

Fig. 4.2.9. Volumes (O) and dry weights (●) for every second leaf of tobacco plotted on equivalent logarithmic scales.

20°/15 °C regime. This accounts for the much shorter plastochrones which may be determined from the initial values of Fig. 4.2.9. Such estimates yield a plastochrone of about 1.3 days between days 7 and 15 and about 1.05 days between days 15 and 27.

Leaf growth

In re-presenting the leaf-growth data of Hannam (1968) I have placed rather more emphasis on those portions of the curves which imply fairly strict exponential growth (Fig. 4.2.9) than did the author herself. This emphasis and the more schematic form of the relative growth rate statement of Fig. 4.2.10 is used to draw attention to the similarities and dissimilarities between this and the other shoot-apical systems described in this chapter. They imply no criticism of the original treatment.

These two text figures should be evaluated together, for the relative growth rates of Fig. 4.2.10 are but the *b* values or slopes of the curves in Fig. 4.2.9. It may be objected that the high initial values rest on only one value for each volume curve – none at all for primordium 2, which has no very early value. However, all of the leaves, not just every

Nicotiana tabacum

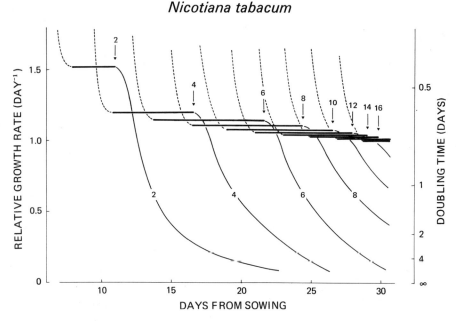

Fig. 4.2.10. Relative growth rates for every second leaf of tobacco. The early, rather speculative portions of the curves are dotted, the exponential phases are shown as heavy horizontal lines, and the fine continuous lines indicate the falling rates as maturation proceeds. The arrows mark the cessation of strictly exponential growth.

second one, which have early values show this feature (Hannam, 1968). Furthermore, high initial values have been demonstrated for flax leaves, and they will be shown or inferred for lupin, *Brassica*, subterranean clover and *Eucalyptus*; so they are credible in tobacco. We also know that each primordium can be regarded as having a pre-initiation rate which is that within the tunica layers of the dome just prior to initiation. Richards's methods of analysis enable us to estimate such initial rates from a knowledge of the plastochrone ratio and the plastochrone interval. They will be twice the radial relative growth rate if the tunica is not increasing in thickness. Such estimates yield values of 0.74 for primordium 6, and 0.59 for primordium 14. It is thus safe to infer that the time course of R_v for the very early growth of tobacco leaf primordia is very similar to that of flax and many other plants. It rises from a low value dictated by the growth rate of the apical dome at origin to a transient peak value before falling back to an intermediate value which is to varying extents governed by restraints imposed by surrounding primordia. That the intermediate value tends to be an exponential one in

90

tobacco is a new phenomenon which will be looked at more fully in due course. It will be suggested that the presence of the masses of leaf hairs may be relevant to that phenomenon.

After varying periods of exponential growth each leaf exhibits the inevitable decline of a maturing organ. However, the leaf does not emerge from the bud until after the decline has set in.

4.3 CAULIFLOWER, *Brassica oleracea* L.

Shoot-apical system

We were introduced to this apical system in Fig. 3.2 where it figures as an extreme example of the Fibonacci spiral type. The following account is based mainly on drawings and volume studies of 22-day seedlings and on a few measurements for 28-day seedlings.

The apical cone on day 22 is rather flat (Fig. 4.3.1D) and the ninth and youngest primordium is already large relative to the residual apical surface (Fig. 4.3.2). Above it is a small tent-like space into which the young primordium can grow freely, but not for long. Primordia 8, 7 and 6 are very much hemmed in by their neighbours and their tips are trapped, at least temporarily, by the encircling margins of the previous primordia. Note in particular the cavity in primordium 5 (Fig. 4.3.1C) from which primordium 6 has been withdrawn (Fig. 4.3.1B). A more complete understanding of the situation can be derived from the comparison of the lower transverse sections of Fig. 4.3.3 with the drawings of the earlier Figure. We may also note, without attempting an explanation, that only the very tip of leaf 5 is trapped within leaf 4. It may be that, as for wheat, later stages of primordium development are less subject to physical constraint than are the early ones described here.

Phyllotaxis

Fig. 4.3.4 presents a transverse projection of this same 22-day apex together with contact parastichy spirals superimposed on the same outlines. The genetic spiral is still very much a contact one; there are two prominent clockwise contact parastichies; and there are also three fairly prominent anticlockwise contact parastichies. Church would have styled it a (1 + 2) apex, but Table 4.3.1, with its phyllotaxis index of 2.0 for day 22, makes it an orthogonal 2:3 parastichy system according to Richards. The Table also shows the plastochrone ratio to fall and the phyllotaxis index to rise with time as in the previous examples of

Brassica oleacea

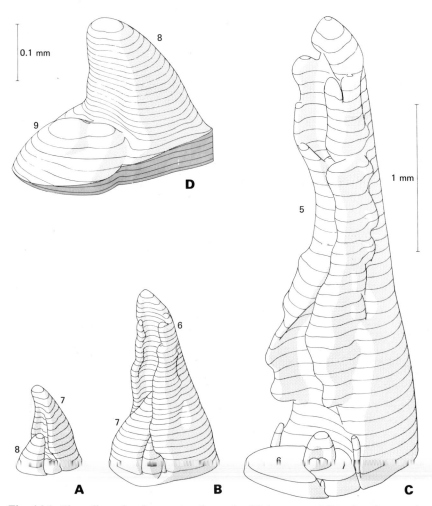

Fig. 4.3.1. Three-dimensional reconstructions of a 22-day apex of *Brassica oleracea*. A. Primordia 7 and 8, with primordium 9 and the apical dome between them. B. Primordium 6 with all younger primordia clasped between its rapidly developing margins. C. Primordium 5 with primordia 6 and 7 'dissected' away to expose the apex and younger primordia. D. Large scale drawing of primordia 8 and 9 and the residual dome. Contour lines in A and B are 40 μm apart, those in C are 80 μm apart, and in D they are 10 μm apart.

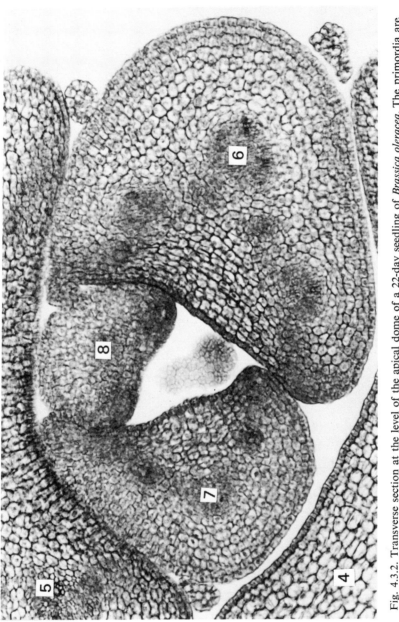

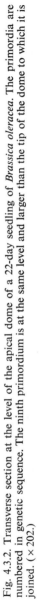

Fig. 4.3.2. Transverse section at the level of the apical dome of a 22-day seedling of *Brassica oleracea*. The primordia are numbered in genetic sequence. The ninth primordium is at the same level and larger than the tip of the dome to which it is joined. (×202.)

Table 4.3.1. *Phyllotactic properties of the apex of*
Brassica oleracea

Age (days)	Plastochrone ratio (r)	Phyllotaxis index (P.I.)	Apex–primordium area ratio	Percentage cover
22	1.633	1.99	0.63	98.2
28	1.519	2.15	—	—

Fibonacci systems. The apex–primordium area ratio is very low (cf. Fig. 4.3.1D) and the percentage cover has the extremely high figure of 98 %. It is so high, in fact, that one is left wondering if the curious vermiform stipules (Fig. 4.3.1B and C), arising when and where they do, could have had any other form.

Leaf growth

In Fig. 4.1.17 curves for the early stages of growth of flax primordia were estimated indirectly by assuming that the volumes of the primordia of an axis could be taken to represent successive values at plastochrone intervals for any primordium of similar age. The same device is used in Fig. 4.3.5 for our 22-day *Brassica* axis, though we are not able here to check its strict validity against a less precise but direct method as was done for flax primordia.

Using volume integration procedures, volumes for the free blade, the buttress and the associated tunica layer were determined for primordia 4–9 inclusive. These were plotted against primordium number (in reverse) either singly or in the combinations indicated in Fig. 4.3.5. The plastochrone was independently shown to be very close to three days for these seedlings, hence the age equivalence scale. A problem which always arises when boundaries between primordium and stem have to be decided is the acceptable thickness of the 'tunica' layer. The unlabelled dotted line above that for the total primordium assumes a tunica thickness twice as great as that regarded as acceptable, and shows what little difference this made to the growth curve.

The most striking feature of Fig. 4.3.5 is the implicit high rate of growth during the plastochrone after initiation. Relative growth rates, R_v, are as follows:

Plastochrone	0	1	2	3	4	5
R_v		0.33	0.94	0.43	0.56	0.60	0.52

Brassica oleacea

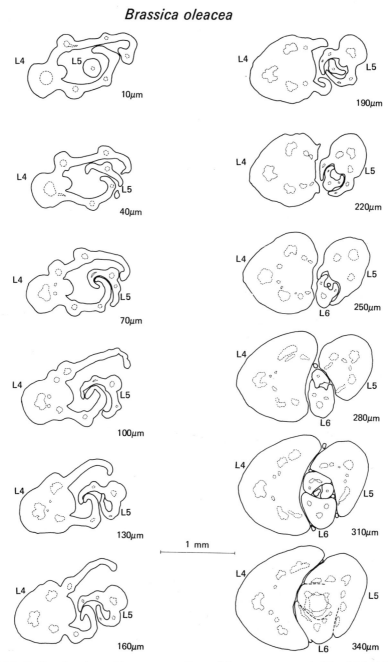

Fig. 4.3.3. Outline drawings of transverse sections of the same apex as in Fig. 4.3.1. All distances are from the tip of leaf 5.

Brassica oleacea

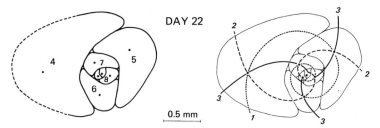

Fig. 4.3.4. Transverse projection and contact parastichies for a 22-day apex of *Brassica oleracea*.

Brassica oleacea

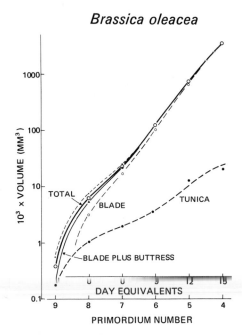

Fig. 4.3.5. Volumes of leaf primordia and their parts as functions of primordium number. For explanation see text.

The value for plastochrone 0 was derived as before from the plasto-chrone ratio of the apex. It is an estimate for the tunica layer just prior to the initiation of primordium 9. Checking these rates against the developmental intervals of Fig. 4.3.1 shows that the high value for plastochrone 1 goes with the unrestrained growth of primordium 9; that the low value for plastochrone 2 goes with the impending restraints

96

to primordium 8; and that the increases in plastochrones 3 and 4 synchronize with the onset of activity in the marginal meristems and the possible easing of physical constraints suggested in §4.3 on shoot-apical systems. It will be noticed that the time course of R_v is essentially the same as that for primordia 8, 24 and 52 of flax (Fig. 4.1.18).

4.4 BLUE LUPIN, *Lupinus angustifolius* L.

Shoot-apical system

This apex was used by the author (Williams, 1970) to illustrate some of the things which can be done with three-dimensional drawings based on serial sections of shoot apical systems. Fig. 4.4.1 is from a single seven-day apex of the blue lupin, and the three separate drawings show the apex 'dissected' in various ways. Taken in conjunction with Fig. 4.4.2 this figure reveals the relationships of successive primordia within the apex and, since adult leaves 6–14 are all heptafoliolate and do not differ greatly in size, gives a satisfactory description of form change in a developing primordium.

The apical dome is a large one, even in a seven-day seedling, and the leaf primordium arises as a fairly localized bulge on its flank. The bulge grows through two plastochrones before a pair of stipules initiate as small peaks subtending an angle of *c.* 90° at the centre and the bulge itself virtually becomes the median foliole (primordium 13). During the next plastochrone the first pair of lateral folioles emerge close to and below the median one, and together they form a central column which is increasingly distinct from the stipules (primordium 12). Then, at unit plastochrone intervals, the second and third pairs of folioles are initiated (primordia 11 and 10), thus completing the full complement of discrete members of the leaf. Later stages, represented by primordia 9, 8, 7 and 6, indicate a close approximation to exponential growth, and the gradual appearance of a basal column of tissue which will become the petiole. The petiole is united with the stipules for a considerable distance above its junction with the stem. Perhaps the most curious feature of this morphogenetic pattern is the production of the folioles progressively in an oblique circle at the top of the emerging petiole.

Phyllotaxis

From the transverse projection of this axis the phyllotaxis is clearly $(2+3)$ by contact parastichies, but it is an almost orthogonal $3:5$ in Richards's terminology. The plastochrone ratio is 1.194 and the plasto-

Lupinus angustifolius

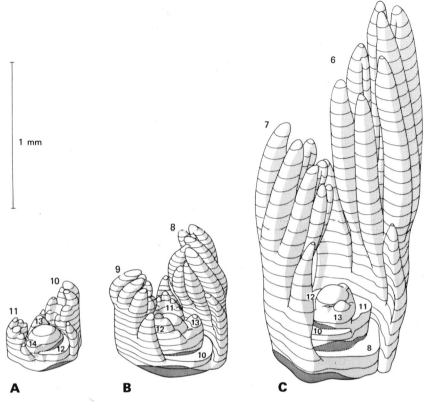

DAY 7

Fig. 4.4.1. Three-dimensional reconstructions of a 7-day apex of the blue lupin. A. The apical dome and primordia 10–14, with the blade of primordium 12 removed. B. The same with primordia 8 and 9 added, and no. 10 removed. C. The same with primordia 6 and 7 added, and primordia 8, 10, 11 and 12 removed. The contours are 40 μm apart in A and B, and 80 μm apart in C. (From Williams, 1970, Fig. 6.)

chrone approximately two days for the conditions of growth. It follows that the radial relative growth rate is 0.089 day^{-1}, and that for the tunica layer prior to primordium initiation is 0.177 day^{-1} (i.e. $\times 2$).

Leaf growth

As for *Brassica* there is no direct information on the early growth of the lupin leaf primordium. However, parallel data to that of Fig. 4.3.5 for *Brassica* is also available for the seven-day lupin seedling, and, since

98

Lupinus angustifolius

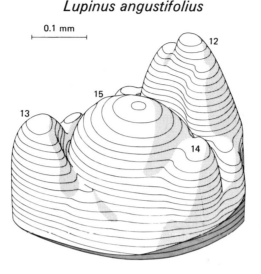

0.1 mm

13 · 15 · 12 · 14

DAY 7

Fig. 4.4.2. The apical dome and very young primordia of the same apex. Primordium 15 on the far side of the dome is just being initiated. The contours are 10 μm apart.

the adult leaves of all the primordia concerned are of similar size, the procedure is believed to give a reliable picture of growth (Fig. 4.4.3). From the total volumes of Fig. 4.4.3, relative growth rates, R_v, are as follows:

Plastochrone	0	1	2	3–4	5–6	7–8	9–11
R_v		0.18	0.68	0.37	0.34	0.35	0.39	0.50

As already noted, the value for plastochrone 0 is that for the tunica layer just prior to initiation of primordium 15; the ephemeral peak for plastochrone 1 goes with the seemingly unrestrained growth of primordium 14; the next three values relate to primordia whose radial growth rate may well be restricted by neighbouring primordia; and the subsequent rise to a value of 0.50 day^{-1} goes with the establishment of marginal meristems on all the folioles. The lupin primordia are also embedded in large hairs, and these could affect growth rates within the apex.

Sunderland *et al.* (1956) obtained a very similar growth curve for the young primordia of *Lupinus albus*. Relative growth rates for roughly comparable plastochrones yield the following results:

Plastochrone	1	2	3	4	5	6
R_v	0.19	0.14	0.15	0.17	0.18	0.18

Lupinus angustifolius

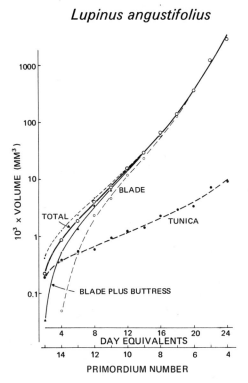

Fig. 4.4.3. Volumes of leaf primordia and their parts as functions of primordium number. For detailed explanation see the text relating to Fig. 4.3.5.

The two sets of data are thus in satisfactory agreement if one allows for the different techniques used, and the necessary omission of tunica volumes from the white lupin study. It is also of interest to note that the forms of these growth curves are essentially the same as those reported above for flax and cauliflower, and also that for tobacco for the early part of its course.

4.5 SUBTERRANEAN CLOVER, *Trifolium subterraneum* L.

Shoot-apical system

The general structure and appearance of the shoot-apical system of sub-terranean clover is shown in Fig. 4.5.1 This structure is similar in all the clovers I have examined, and may be taken to represent the condition in many trifoliate legumes. The third leaf has been dissected away in A, revealing the axillary bud of leaf 3, the whole of leaf 4, and the tips of the folioles of leaf 5. Indeed, leaf 5 is almost completely hidden by the stipules of leaf 4. In Fig. 4.5.1B the same axis is shown with the near halves of leaves 4, 5 and 6 dissected away, so as better to show the inter-relations of successive leaf primordia above the shoot apex. Primordia 7 and 8 are clearly visible, with the apical dome between them. It will be noted that the midribs of the primordia are not co-planar. A note-worthy property of this system is that successive leaves follow each other in what can best be described as a developmental tunnel, and, if leaf 3 could have been included to the same scale, it would be even more evident that the system was made up of an exponential series of similar forms.

Events at the shoot apex itself are depicted in more detail in Fig. 4.5.2. The four stages cover the initiation of primordium 5 and the very early growth of primordium 4. Each primordium arises as a bulge quite high up on the flank of the apical dome and quickly spreads around the dome. Growth in a radial direction would appear to be restricted by the presence of older primordia, so that axial growth quickly becomes the dominant component. The tip of the primordium produces the median foliole and the lateral folioles arise synchronously with the next leaf primordium (between day 5.2 and 5.9 of Fig. 4.5.2). The stipular primordia seem to arise almost immediately thereafter; they are not delayed another plastochrone as might have been expected. It will be evident that the growth rate within the primordium as a whole is considerably greater than it is in the residual tissue of the dome.

Change in form is both rapid and dramatic during these first few days of primordium growth. Some features of subsequent changes in form are deducible from Fig. 4.5.1, but they are more clearly seen in Fig. 5.1 of the next chapter. This illustrates the changing size and form of the fourth leaf at two-day intervals from day 7 to day 13. The drawing for day 7 is, of course, a much reduced version of that for day 7 in Fig. 4.5.2. By day 7 the marginal growth initiating the leaf blades had commenced, and growth of the blades is a prominent feature for several days. For *T. repens* Denne (1966b) found marginal activity to be short-lived, and there was no evidence that cells at the margin divided more rapidly than those of the rest of the lamina. This may be presumed to hold for *T. subterraneum* also. The petiole arises as a constriction below the three folioles on day 9, and soon establishes the columnar form of days 11 and 13. In the meantime the stipular initials surround the axis and quickly assume the adult form. After day 11 there are no major changes of form, though the petiolules are to be characterized before leaf emergence on

101

Trifolium subterraneum

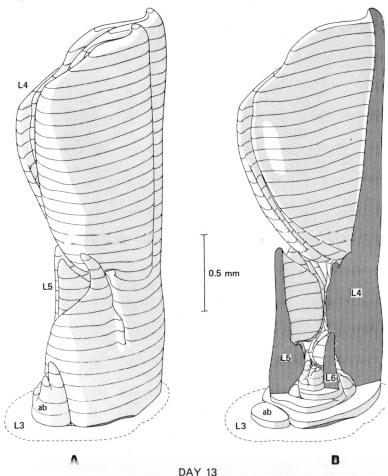

DAY 13

Fig. 4.5.1. Three-dimensional reconstructions of a 13-day apex of subterranean clover. A. As seen after removal of leaf 3 at its point of half-junction with the axis. B. As seen after the removal of the axillary bud (ab) and the near halves of leaves 4, 5 and 6. Contour lines are 80 μm apart except near the apical dome where they are 40 μm apart.

or about day 16. Fig. 4.5.6 shows that length growth is exponential from initiation to emergence (days 5–15 for the fourth leaf), so a three- to four-fold increase in length beyond that illustrated in Fig. 5.1 takes place before that event. Thereafter, of course, the petiole continues to elongate and the leaf blades spread as full maturity approaches.

102

Trifolium subterraneum

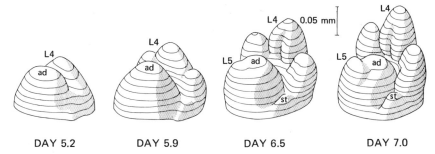

DAY 5.2	DAY 5.9	DAY 6.5	DAY 7.0

Fig. 4.5.2. Three-dimensional drawings representing form changes near the apical dome during a single plastochrone. Age equivalents are shown to the nearest tenth of a day, and the contour lines are 10 μm apart: ad, apical dome; st, stipule. (Williams and Bouma, 1970, Fig. 4.)

Phyllotaxis

Having described the shoot apical system in some detail, with special attention to the genesis of form, we now turn to its phyllotactic properties and their quantitative description. Outline drawings A, B and C of Fig. 4.5.4 are from selected serial sections through a 13-day apex of subterranean clover. They and the photomicrograph of Fig. 4.5.3 illustrate once again the beautiful packing of parts within the bud, and the spatial relations of the successive foliar members. Fig. 4.5.4D is the transverse projection or plan of the areas of attachment of successive primordia, and E is a contour plan of leaf primordium 7 and the apical dome. The centre of the phyllotactic diagram, F, was determined by trial and error so as to give a uniform progression for the distances of the 'centres' of the successive primordia. The plastochrone ratios of Table 4.5.1 are the ratios of the successive members of Fig. 4.5.4F and it is clear that the characteristic phyllotaxis for the species was still being established. The divergence angle was very high between primordia 3 and 4, but was stabilizing at about 160° thereafter. From an analysis of six successive divergence angles in each of six seedlings (ages between 22 and 29 days) no effects of plant age or leaf number could be established. The general mean was 165.08° ± 1.42° ($n = 36$). There was no evidence to suggest that the divergence angle was closer to the Fibonacci angle to begin with.

In order to determine the characteristics of the system with some precision, further measurements were made on the material studied by Williams and Bouma (1970). Plastochrone ratios were determined for

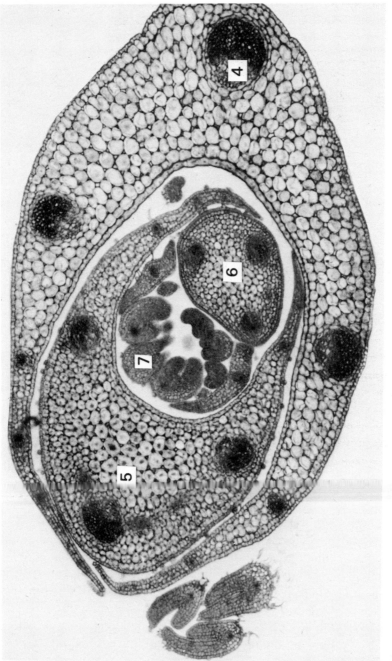

Fig. 4.5.3. Transverse section at the level of the tip of the apex of a 20-day plant of *T. subterraneum*. The primordia are numbered in their order of formation. (×105.)

Trifolium subterraneum

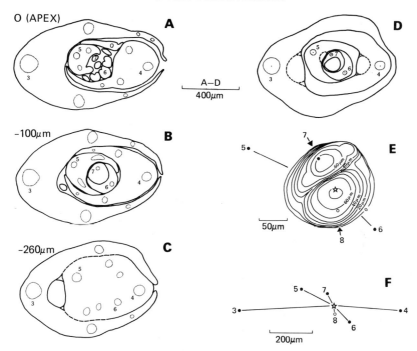

Fig. 4.5.4. A. Transverse section through the stem-clasping bases of leaves 3 and 4, the petiole and stipules of leaf 5, trifoliate leaf 6, and the tips of leaf 7 and the growing point.

B. Transverse section 100 μm below A, and through the stem clasping bases of leaves 3–6, and the stem just below its junction with leaf 7.

C. Transverse section 260 μm below A, and through the stem at the point of half-junction with leaf 3. The bud in the axil of leaf 3 is a prominent feature.

D. Composite plan of leaf attachment areas built up from projections of all available sections from above A to below C. Only the mid-bundles are shown, and their positions are established at the points of junction of the primordia with the stem.

E. Enlarged contour plan of leaf 7 and the growing point. The black spots mark the centres of the mid-bundles of leaves 5, 6 and 7; the star marks the centre of the system; and the small circle is the predicted centre of leaf 8.

F. Phyllotactic diagram of the system (see Table 4.5.1 for numerical evaluation). The outlines of provascular strands and early stage vascular bundles are shown in A–C. (Williams and Bouma, 1970, Fig. 3.)

plants ranging in age from 1 to 29 days and having up to seven primordia available for measurement. A total of 56 values were so obtained. The ratios tended to be high close to the dome but decreased with increasing distance. It is also true that the ratios for a given leaf interval decreased with time. The results are summarized in Table 4.5.2, from which it can readily be seen that the decreases were highly significant. It takes rather little imagination to realize that the decline is inevitable in a system in

105

Table 4.5.1. *Phyllotactic properties of the apex shown in Fig. 4.5.4*

Between primordia	Divergence angle	Plastochrone ratio (r)
3 and 4	173°	1.42
4 and 5	157°	1.85
5 and 6	162°	1.71
6 and 7	161°	1.61
Mean	163°	1.65

Table 4.5.2. *Plastochrone ratio in* T. subterraneum *as a function of distance from the centre of the apical dome*

Upper radial distance	Plastochrone ratio (r)	S.E.	n
< 300 μm	1.735	± 0.028	28
	***		
Between 300 and 600 μm	1.539	± 0.032	14
	*		
> 600 μm	1.442	± 0.028	14

$* P < 0.05$, $*** P < 0.001$.

which radial growth of the stem is so limited; the stems were less than 2 mm in diameter in this instance. This exercise has clearly demonstrated the necessity to work close to the apex, even though large numbers must be measured to obtain reasonable precision. The best estimate for the Mount Barker cultivar of *T. subterraneum* is thus 1.735 ± 0.028, and this is equivalent to a phyllotaxis index of 1.87. The corresponding value for the apex–primordium area ratio is 0.664 ± 0.016, which simply tells us that, at the time of leaf initiation, the projected area of the residual apex is only 66 per cent of that of the primordium which has just been formed.

This apical system has much in common with the spirodistichous system of phyllotaxis found in several monocotyledonous families (Snow, 1951). This is usually derived from distichy during ontogeny, the first leaf of the seedling being opposite the cotyledon. After the change to spirodistichy, the leaves are laid down with a divergence angle of less than 180° but greater than the Fibonacci angle of 137.5°. The mean divergence angle for *Rheo discolor* (from Fig. 1 of M. Snow, 1951) is in fact 154°, and is thus closer to the Fibonacci angle than is that for *T. subterraneum*.

A characteristic of spirodistichy is that each young primordium covers an arc of more than 180° before the next leaf arises, and in some species each leaf may encircle the apex before this happens. That this condition is approached in *T. subterraneum* is clear from Fig. 4.5.2, where the stem-clasping stipules of the leaf are seen to arise precociously and to yield stem-circling attachment areas (Fig. 4.5.4). A further point of agreement lies in the tendency to asymmetry in the lateral parts of the attachment areas in *Trifolium*. However, the similarity may end there, for the phyllotactic system in *Trifolium* derives from the decussate condition of the cotyledons, via the distichy of the first two or three foliage leaves. These are, of course, at 90° to the cotyledons, and the divergence angle is close to 180°. Even between leaves 3 and 4 (Fig. 4.5.4) the angle is 173°, but quickly approximates to 165° thereafter. R. Snow (1965) has investigated similar large divergence angles between leaves in Cucurbitaceae. From illustrations provided by Hagerup (1930), Snow calculated divergence angles of 146.6°, 157° and 164.7° for *Ecballium elaterium*, *Lufa cylindrica* and *Thaladiantha dubia*, respectively. In *Cucurbita pepo*, with which Snow himself worked, the mean divergence angle was 153°. Since all these angles are said to be 'abnormally large', it is clear that they are thought to derive from the spiral phyllotaxis commonly encountered in dicotyledons. Snow concludes that the large divergence angles are referable to the eccentricity of the precocious axillary buds. However, whatever the cause of aberrancy in specific cases, all these apical systems have similar numerical properties. It seems logical, therefore, to describe them all as spirodistichous.

The apex proper

It is appropriate at this point to consider the elegant study by Denne (1966a) of the shoot apex of *Trifolium repens*. The apex was defined by her as the growing point above the youngest primordium, including the 'shell' zone in the axil of the penultimate primordium (Fig. 4.5.5). Five regions of the apex were recognized: S, the summit region, including some corpus cells; F_1, flank cells of the next primordium – four layers deep, and with periclinal divisions appearing in the third or fourth layers; F_2, flank cells of the subsequent primordium; R, the rib meristem; and A, the axillary region, or 'shell' zone. F_2 includes its associated A region. Although it was not possible to draw precise boundaries between the regions, they were adequate for estimates of cell number, cell size and mitotic index. Denne found that the mitotic indices of these regions were consistent with their rates of cell division as measured by accumulation of metaphases after colchicine treatment. The data of

Trifolium repens

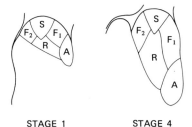

STAGE 1   STAGE 4

Fig. 4.5.5. Outline diagrams of median longitudinal sections of *Trifolium repens* at early and late stages of the plastochrone. Definitions of the regions are in the text. (After Denne, 1966*a*, Fig. 1.)

Table 4.5.3 are derived from the more voluminous data of Table 1 of the original paper, where it was shown that the mitotic index was significantly higher in flank regions of the apex than at the summit, and higher in the summit than in the rib or axillary regions. There was little variation of duration of mitosis between regions.

Changes in these attributes during the plastochrone were also determined, though the results should perhaps be accepted with caution. There is a suggestion that the mitotic index declined from 4 to 3 in F_1 and rose from 2 to 3 in F_2, thus implying a cyclic fluctuation in cell division of the flank regions during the plastochrone. This is consistent with the transient high values of R_v during primordium initiation reported elsewhere in this book for subterranean clover, flax, tobacco, and some other dicotyledons.

Leaf growth

Having described the apical system in some detail, it is time to examine the growth rate characteristics of the succession of foliar members which are initiated and grow out from the apex proper. Length growth data are the most readily obtained, and it has already been noted that axial, or length growth quickly becomes the dominant component, possibly because of the constraints imposed by the presence of older primordia. The length data for successive leaves of the main shoot of subterranean clover grown in a controlled environment (Williams and Bouma, 1970) are presented in Fig. 4.5.6. Apart from the day 1 values for leaves 1 and 2, all values up to a length of 10–12 mm are well described by a family of linear regressions. The procedure adopted for

108

Table 4.5.3. *Variation of mitotic index, duration of mitosis and duration of mitotic cycle within the apex of* T. repens L. (*Based on Denne, 1966a*)

Attribute	Region of apex				
	S	F_1	F_2	R	A
Mitotic index*	2.33	3.71	2.63	1.84	1.46
Ratio	*1.00*	*1.60*	*1.13*	*0.79*	*0.63*
Duration of mitosis** (h)	4.1	4.0	3.7	3.3	4.2
Duration of mitosis cycle** (h)	108	69	87	137	212

* Cells in late prophase or early telophase as percentage of all cells present in the region. Weighted means based on 921 apices from 4 experiments.
** Calculated 4 and 12 h after colchicine treatment of 32 apices.

establishing such a family is described in the appendix. That the values for day 1 do not fit into this scheme may only mean that growth was re-established slowly after dormancy. That the regressions are linear for a logarithmic plot of the data implies, of course, that length growth is exponential so long as the leaves remain in the bud.

A striking feature of Fig. 4.5.6 is the fall in regression slope with leaf number. This is equivalent to a decline in length relative growth rate R_l from 0.566 to 0.352 day^{-1} for leaves 1 and 14 respectively. Put in another way, leaf 1 took only 1.22 days to double its length, whereas leaf 14 took 1.97 days to do so. Following emergence, R_l declines progressively to zero over a period of at least a week.

From Fig. 4.5.2 it will be evident that the direct determination of times of initiation of successive primordia and hence of the plastochrone interval would be difficult. They can be indirectly estimated from Fig. 4.5.6 where the regressions have been extrapolated to the points in time at which the primordia have a length of 50 μm. Such a length is a little in excess of the thickness of the disc of insertion at its earliest measurable stage (cf. leaf 5, day 6.5 of Fig. 4.5.2), and is a reasonable threshold value for leaf length. From the whole series it can be shown that the plastochrone decreases to a minimum of 1.5 days for leaf interval 6–7, then increases slowly to 2.0 days for 13–14. On the other hand, the plastochrone for leaf emergence (taken as a length of 10 mm) averages 2.2 days between leaves 1 and 10. That this is in excess of the

Trifolium subterraneum

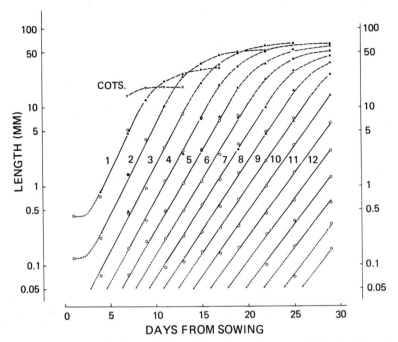

Fig. 4.5.6. Length data for the cotyledons ($\triangledown$) and successive leaves 1, 2, etc. of the primary shoot. The solid triangles ($\blacktriangle$), are from macroscopic measurements, but the circles ($\bigcirc$) are values derived from serially sectioned material. (Williams and Bouma, 1970, Fig. 5.)

corresponding mean plastochrone for initiation (about 1.6 days) follows from the trend in regression slopes.

Williams and Bouma (1970) tabulate dry weights and leaf areas for the same set of leaves for which the lengths are portrayed in Fig. 4.5.6. However, for our present purposes, we can confine our attentions to the fresh weights. Values for leaves 1–11 are plotted on an absolute scale in Fig. 4.5.7 because it is salutary to be reminded of the shapes and properties of absolute growth curves in a book which is dominated by the consideration of relative rates of change. The curves indicate a progressive increase in final size at least to leaf 4; it is unlikely that later leaves would have been much larger. The curves have a non-symmetrical, sigmoid form which is best described by the flexible curves of Richards (1959).

The parallelism between change in volume and fresh weight has been used in assembling Fig. 4.5.8. Eight common values were used to link

Trifolium subterraneum

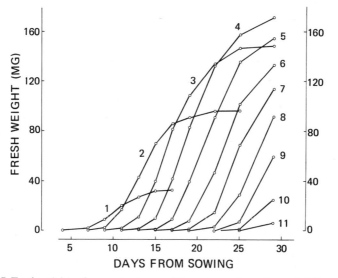

Fig. 4.5.7. Fresh weights of successive leaves of the primary shoot (leaves 1–11) as a function of time. (Williams and Bouma, 1970, Fig. 8.)

the volume and fresh weight scales, and their mean values imply an apparent density of 1.66 g cm^{-3}. This is higher than the true density because of the appreciable though fairly consistent shrinkage of the tissues during processing.

The most striking feature of Fig. 4.5.8 is that so much of its information can be accounted for by the array of linear regressions for values ranging from 10^{-3} to 8 mm^3. The changes in slope of these regressions with leaf number – the R_v values – are shown in Fig. 4.5.11, and in both magnitude and duration as the heavy horizontal lines of Fig. 4.5.9. R_v values for the first and the fourteenth primordia are 1.29 and 0.77 day^{-1} respectively, and imply doubling times of 0.54 and 0.90 day. The form of the growth curves subsequent to the periods of exponential growth are those to be expected for leaves undergoing the normal processes of maturation. It should be remembered that more than 90 % of the substance of the leaf is laid down during maturation – the periods covered by the broken lines in Fig. 4.5.8 and the fine continuous lines of Fig. 4.5.9.

The pattern of volume growth in the very young leaf primordia (volumes ranging from 2 × 10^{-5} to 10^{-3} mm^3 and indicated by the dotted lines of Fig. 4.5.8) is established with much less certainty, but is

Trifolium subterraneum

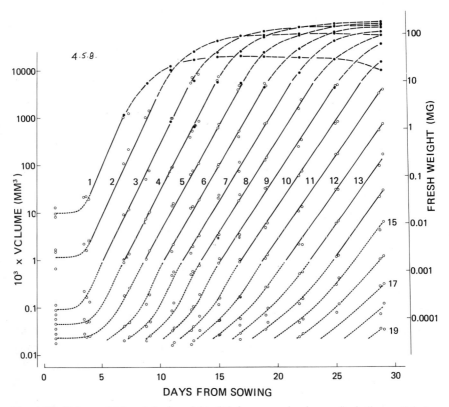

Fig. 4.5.8. Volumes (○) and fresh weights (●) for successive leaves (including petioles) plotted on equivalent logarithmic scales. The fresh weight scale is extrapolated downwards to provide estimates based on an assumed constancy of weight per unit volume. (Williams and Bouma, 1970, Fig. 9.)

believed to be sound. For instance, leaves 6 to 10 all have experimental values which are below the negative extrapolations of their phases of exponential growth. This implies brief periods of high relative growth rate such as are shown for these leaves in Fig. 4.5.9. There is no evidence for this feature in primordia 1–4 or 12–19.

Fig. 4.5.9 provides an effective summary of the pattern of growth within the vegetative apex, and doubling times can be read off from the right-hand scale. A curve drawn through the initial values for R_v in this Figure provides estimates of growth rates on the flanks of the apical dome. A fuller understanding of the pattern is helped by studying the three-dimensional drawings (Figs. 4.5.1, 4.5.2 and 5.1) and the dia-

112

Trifolium subterraneum

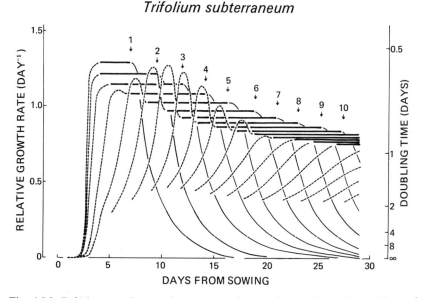

Fig. 4.5.9. Relative growth rates for successive leaves of the primary shoot. The early, rather speculative portions of the curves are dotted, the exponential phases are shown as heavy horizontal lines, and the fine continuous lines mark the falling rates as maturation proceeds. The arrows mark the cessation of strictly exponential growth, and this occurred when each leaf was 5–10 mm long. (Williams and Bouma, 1970, Fig. 12.)

grammatic longitudinal sections of Fig. 4.5.10. These all remind one of the analogy already mentioned – that of a developmental tunnel in which successive members of an exponential series of similar forms follow each other. Fig. 4.5.10 makes the additional point that these forms are more tightly packed on day 29 than on day 11. This could be relevant to the declining values of R_v for the exponential phases of growth and to the disappearance of the initial maxima.

Williams and Bouma (1970) also present arrays of linear regressions which describe the volume growth of the corpus tissue associated with successive leaf primordia, and for their axillary buds (or runners). Relative growth rates for all three entities during their phases of exponential growth are plotted against leaf number in Fig. 4.5.11; R_v values for corpus tissue are approximately half of those for the leaf primordia and fall from 0.71 to 0.36 day^{-1}. Apart from the first few values in the series for the runner buds, and particularly in the first, R_v is only a little less than for the associated leaf. That the first value is low could be related to the fact that its site is present in the seed, and bud initiation is somewhat inhibited.

113

Trifolium subterraneum

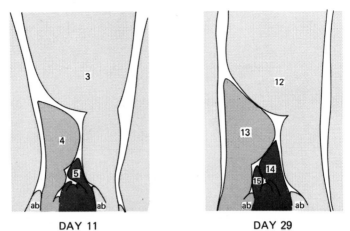

DAY 11   DAY 29

Fig. 4.5.10. Diagrammatic longitudinal sections reconstructed from serial transverse sections of apices of the ages indicated. The intensity of stippling is a rough guide to the intensity of meristematic activity. The primordia are numbered in order of appearance: ab, axillary bud.

Trifolium subterraneum

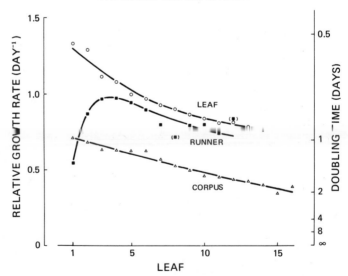

Fig. 4.5.11. Relative growth rates for the exponential phases of the growth of leaf primordia, associated corpus tissue, and axillary buds (runners) as a function of leaf number. (Williams and Bouma, 1970, Fig. 13.)

114

Trifolium subterraneum

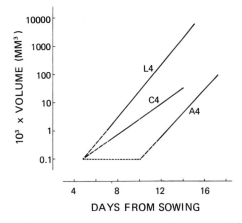

Fig. 4.5.12. Diagram illustrating the time relations of the exponential phases of growth of the leaf primordium (L), the corpus tissue (C), and axillary bud (A) of the fourth leaf. (Williams and Bouma, 1970, Fig. 14.)

This extensive set of data provided an indirect but objective approach to the determination of times of initiation of leaf primordia and axillary buds, and of the lapse of time between these events. The approach rests on the fact that the growth of all parts is exponential for long periods. The procedure is illustrated for the fourth leaf in Fig. 4.5.12 where the regression lines for the leaf primordium (L4), the corpus tissue (C4) and the axillary bud (A4) are plotted together. If L4 and C4 are extrapolated they intersect at a point where corpus and primordium volumes are equal, and this is about the time at which the primordium is initiated. If the volume at this time be accepted also as reasonable for an axillary bud at initiation, then its intercept with A4 will give both the time of its initiation and the required lapse of time (Fig. 4.5.12). The results of this exercise are recorded in Table 4.5.4 for primordia 1–12. The first point to note is that the common volumes are fairly uniform, and that the mean corresponds to a volume of 7.7×10^{-5} mm^3. Such a volume would be made up of rather more than 300 cells on the flank of the apical dome. The time lapse between leaf and bud initiation increased from five to seven days with increasing leaf number. This kind of analysis could conceivably be of value for studies of environmental and nutritional effects on growth and lateral shoot production.

Table 4.5.4. *Interrelations of the exponential phases of growth of leaf primordium (L), corpus (C), and axillary bud (A) of successive leaves (Fig. 4.5.12)*

Leaf no.	Intersection of $\log_{10}V$ for L and C		Time of initiation of A	
	Time (days)	$10^3 \times$ volume (mm³)	Time (days)	Days after initiation of L
1	−0.43	0.091	4.98	5.41
2	1.24	0.093	6.11	4.87
3	2.96	0.094	8.08	5.12
4	4.76	0.098	9.99	5.23
5	6.43	0.100	11.88	5.45
6	7.80	0.093	13.49	5.68
7	9.08	0.086	14.99	5.91
8	10.31	0.079	16.64	6.33
9	11.60	0.072	18.18	6.58
10	13.37	0.077	20.26	6.89
11	15.33	0.082	22.26	6.93
12	17.88	0.099	24.76	6.88
Mean		*0.089*		*5.94*

4.6 EUCALYPTUS, *Eucalyptus grandis* Hill and Maiden and *Eucalyptus bicostata* Maiden, Blakely and Simmonds

Shoot-apical system

The genus *Eucalyptus* was selected as a representative of the opposite, decussate system of phyllotaxis for a number of reasons other than its obvious importance in Australia. Unlike most decussate plants, *Eucalyptus* does not remain decussate during vegetative growth but escapes into a condition which is usually described as alternate, but which is really a simple modification of the decussate condition. *E. grandis* was chosen because this change takes place quite rapidly and smoothly during early seedling growth. It is seen in Fig. 4.6.1C, especially in the younger internodes near the terminal bud, and perhaps more clearly in Fig. 4.6.3B. In that Figure, all but the very small primordia surrounding the apex are dissected off horizontally at their points of half-junction with the axis, but leaving the axillary buds intact. The buttresses of 8 and 8′, 9 and 9′, 10 and 10′ are opposite but at different levels on the stem, and each pair is at 90° to the previous one. That the members of each pair are at different levels is a secondary phenomenon, developmentally speaking, hence my preference for modified decussate rather than

Fig. 4.6.1. A–C. Seedling growth of *E. grandis*; A – 15 days from sowing, B – 27 days, C – much later, showing modification of decussate leaf arrangement and precocious development of axillary buds. D–F. Seedling growth of *E. bicostata*; D – 15 days from sowing, E – 27 days, F – much later showing strict decussate leaf arrangement and square stems with prominently winged edges. (A, B, D and E, ×1.5; C, ×0.8; F, ×0.5.)

alternate as the appropriate description of the system. However, mature lateral shoots often achieve a pseudo-alternate condition by secondary twisting of petioles or stems.

In *E. grandis* there is no sharp distinction between juvenile and mature leaves; there is only a progressive increase in size and a slow change in shape. In other eucalypts, well represented by *E. bicostata*, the juvenile leaves are opposite, stem-clasping and dorsiventral (Fig. 4.6.1F), whereas the mature leaves are pseudo-alternate (decussate by origin), petiolate and isobilateral. The transition from one type to the other tends to be abrupt, seems to happen between growing seasons, and the types are separated by a few bract-like leaves which readily abscise. The two eucalypts examined here are thus representative of the two main types of heteroblastic development discussed by Allsopp (1967), and the choice was originally made in the hope of shedding some light on the nature of the differences between them.

Fig. 4.6.2 is based on a 60-day seedling of *E. bicostata*, so dissected as to reveal the relationships of the leaf pairs near the apex. Each leaf pair is about half the length of its predecessor, though the youngest pair (not visible here) is always very much less than half the length of the penultimate pair. Axillary buds are rather precocious, being normally present in the axils of the third pair of primordia below the apex. Young stems are rectangular in section and they have prominent wings (see also Fig. 4.6.1F) which run on to the sides of the leaf buttresses, but are not continuous with the leaf blades. Comparison with Fig. 4.6.3 shows that, even in the juvenile condition, *E. grandis* has no wings on its rectangular stem, the leaves have distinct petioles, and never become stem-clasping.

Both these apical systems set rather special problems when it comes to the determination of volumes, and the solutions adopted are illustrated in Fig. 4.6.3C for *E. grandis*, and in Fig. 4.6.4 for *E. bicostata*. The most objective reference points are the junctions of the primordia with the axis, and these rather neatly define the rectangular limits to the 'internodes' of the leaf pairs. The details of the procedure will be considered again in an appendix; the diagrams are given here as an aid to understanding the systems.

We can now look more closely at the apex of *E. bicostata* as this is illustrated in Figs. 4.6.5 and 4.6.6. On day 87 the eleventh pair of leaf primordia are only 0.5 mm long, their tips are recurved, a considerable part of their adaxial surfaces are in contact, and they almost completely enclose the twelfth pair of primordia (Fig. 4.6.5A). The twelfth pair is

118

Eucalyptus bicostata

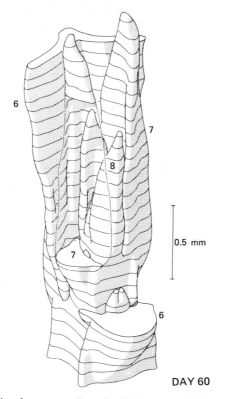

0.5 mm

DAY 60

Fig. 4.6.2. Three-dimensional reconstruction of a 60-day apex of *E. bicostata* dissected to show the relations of the younger pairs of leaves.

only 0.1 mm long (Fig. 4.6.5B) and their tips are together above the apex proper, which is reduced to a narrow saddle-like area of tissue between them (Fig. 4.6.5C). This drawing suggests the presence of incipient bulges at the presumed sites of the thirteenth pair, but their active initiation must clearly await the attainment of a sufficient area of bare apex. Fig. 4.6.6 illustrates an early stage in the development of the tenth pair of primordia – a stage through which all pairs may be supposed to pass. An important point to note is that the form of the twin primordia is rapidly distorted between this stage and that represented in Fig. 4.6.5B by the growth of the marginal meristems of the previous leaf pair. The same sequence of events is illustrated in B, C and D of Fig. 4.6.7. In B the section passes through the apical saddle and the eleventh pair of primordia – indeed some early periclinal

119

Eucalyptus grandis

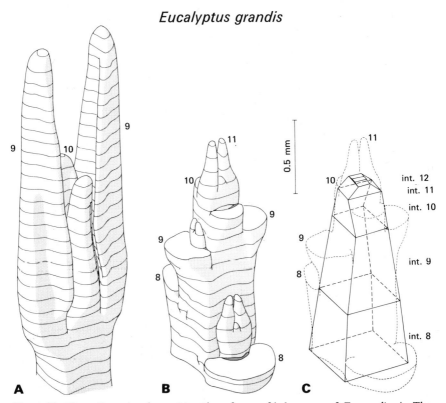

Fig. 4.6.3. Three-dimensional reconstructions from a 91-day apex of *E. grandis*. A. The ninth leaf pair and above shown intact. B. The axis from below the eighth pair, with leaf blades dissected away to illustrate the modified decussate system. C. Skeleton diagram defining the limits of leaf buttresses and 'internodes'.

Eucalyptus bicostata

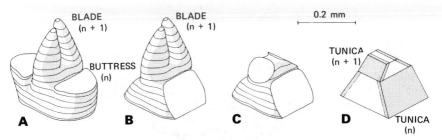

Fig. 4.6.4. Three-dimensional reconstruction for *E. bicostata* showing the implications of the procedures adopted for defining the limits of leaf blade, buttress, tunica and corpus. A. With the blades of leaf pair n removed. B. With their buttresses removed. C. With blades and buttresses of pair n+1 removed. D. The geometrical equivalent of the two upper 'internodes'. The tunica layers are indicated by stippling.

divisions relating to the thirteenth pair are to be seen in the right-hand shoulder of the saddle (cf. Fig. 4.6.5, which is based on an axis at an almost identical stage of development). That the tips of the twelfth pair had already been forced together is evidenced by the presence of scrapings of the adaxial surfaces from both members in the expected position above the apex. Fig. 4.6.7C is for an apex of the same age, but slightly less advanced. It is at right angles to the previous section and so passes through the tenth and twelfth pairs of primordia, with scrapings from the eleventh pair between them. Clearly the tips of the twelfth pair are still well separated in this specimen. Fig. 4.6.7D is from a younger axis and the section is in the plane of the seventh and ninth pairs of primordia. However there is no reason to question that it provides a sound picture of a later stage from that shown in C. On this assumption, the tenth pair of primordia will grow further apart, the twelfth pair will be forced together by the margins of the eleventh pair (actually the eighth in D), and the apex will reconstitute between the twelfth pair of primordia. To describe this last change as a reconstitution is scarcely too strong a way to describe the change from the picture in C to that in D, though the residual apex of C is continuous with that part of itself – in the plane at 90° to C – which is about to initiate the thirteenth pair of primordia.

Fig. 4.6.7 also illustrates stages in the development of axillary bud meristems – especially in B, C and D, though A can perhaps be taken as intermediate between C and D in this respect. Close similarities in the physical surroundings of the apical and axillary meristems is evident in D.

The foregoing example of the probable relevance of physical constraint to the genesis of form in young primordia raises once more its possible relevance to the growth rates of those primordia. At this point, however, I will content myself with establishing the times at which such constraint could operate. For the duration of one plastochrone, or thereabouts, the youngest pair of primordia can grow freely without contact with their predecessors. Fig. 4.6.8 shows for a 'juvenile' type of shoot – as distinct from the seedling – that as many as five pairs of leaves may grow for a time in close contact, for although the drawing shows small gaps between the inner leaves, these are probably artifacts of processing. The larger gaps between pairs 1 and 2 are certainly a natural phenomenon. At the time of collection, the 'mature' apex of Fig. 4.6.8 had its leaves so tightly gummed together that it was difficult to visualize the bud being able to grow except as a unit.

Eucalyptus bicostata

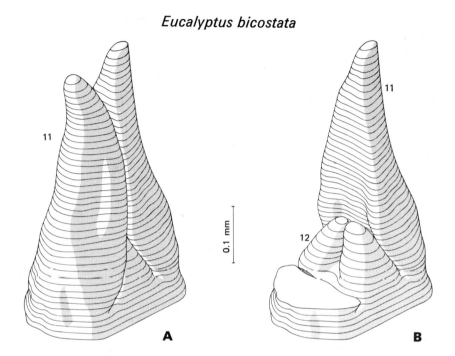

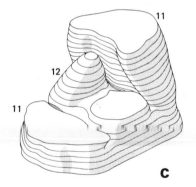

DAY 87

Fig. 4.6.5. Three-dimensional reconstructions of an 87-day apex of *E. bicostata*. A. External view of the whole apex above the junction of the tenth pair of leaf primordia. B. The same with the nearer member of the eleventh pair removed to show the twelfth pair of leaf primordia. C. Further dissected to show the saddle-shaped residual apex.

Eucalyptus bicostata

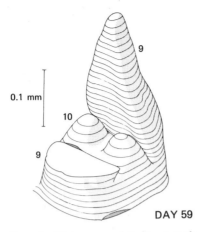

Fig. 4.6.6. Reconstruction of a 59-day apex of *E. bicostata* showing a very early stage for the tenth pair of leaf primordia. Their tips have yet to be forced together by the growth of the marginal meristems of the ninth pair of primordia.

How, then, do these conditions develop? For, in the very young seedling, the first few pairs of leaves obviously grow quite independently. Fig. 4.6.9 attempts to answer this question, though perhaps in a rather clumsy way. The base line of transverse-section outlines shows that the degree of close packing of primordia around the apex is slowly increasing with seedling age. From the vertical columns of outlines it is also possible to assert that this packing extends further above the apex proper in the older seedlings. In examining these drawings it is necessary to try to allow for the possibility that the outer pairs may have lost contact with the inner ones during processing. Thus, in the 104-day apex, the L12 pair would almost certainly have been in contact at the 10 and 50 μm levels, but doubtfully so at the 160 μm level. This apex has only one fewer pairs of leaves in close contact than has the 'juvenile' apex of Fig. 4.6.8.

Phyllotaxis

When discussing Fig. 3.2D above, it was asserted that it is difficult to construct a transverse projection for the apex of *Eucalyptus*. This difficulty arises because of precocious stem elongation and the nature of the leaf buttresses (Figs. 4.6.2 and 4.6.3). Any phyllotactic pattern which exists to begin with is quickly distorted by secondary events.

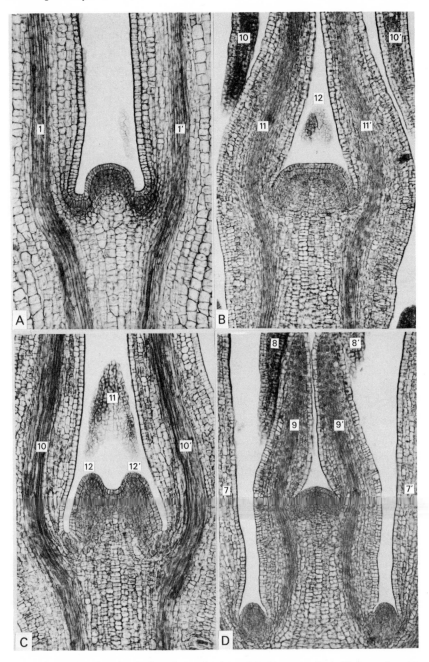

Fig. 4.6.7. Medium longitudinal sections of apices of *E. bicostata*. A. A 13-day axis cut in the plane of the first leaf pair. B. An 83-day axis cut in the plane of the eleventh pair of leaves. C. An 83-day axis cut in the plane of the tenth and twelfth pairs of leaves. D. A 58-day axis cut in the plane of the seventh and ninth pairs of leaves. (× 150.)

Eucalyptus bicostata

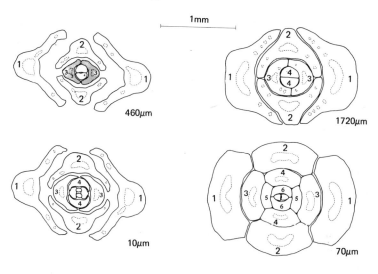

Fig. 4.6.8. Transverse section outlines for 'juvenile' and 'mature' bud types in *E. bicostata*. The measurements in μm refer to distances above the apical dome. That the outer pairs of leaves are labelled 1 in all cases means only that they are the oldest leaves present.

However, a satisfactory transverse projection can be developed for the two or three pairs of primordia near the apex. Stages in the preparation of such a projection are given in Fig. 4.6.10, where the key construction lines – accentuated in the drawings – are the junctions between primordium and axis. Each junction normally takes place within the limits of a single 10 μm section, and, since they are straight, the end result is a combination of rectangles. Not all decussate systems yield rectangular projections, however, for crescent-shaped primordia can produce a pattern equally well. Two ideal projections illustrating this point appear in Fig. 4.6.11. Both are based on the 'golden' section (the irrational factor 0.618034...), such that comparable dimensions of the similar figures are so related. While there is as yet no evidence to suggest that either types are common in nature, a survey might show that they represent optimal solutions to developmental, or space-filling problems.

Returning to Fig. 4.6.10, it remains to define the system by appropriate measurement. Since the vascular bundles are ill-defined at this stage, the most suitable reference points are those half-way between the

125

126

Eucalyptus bicostata

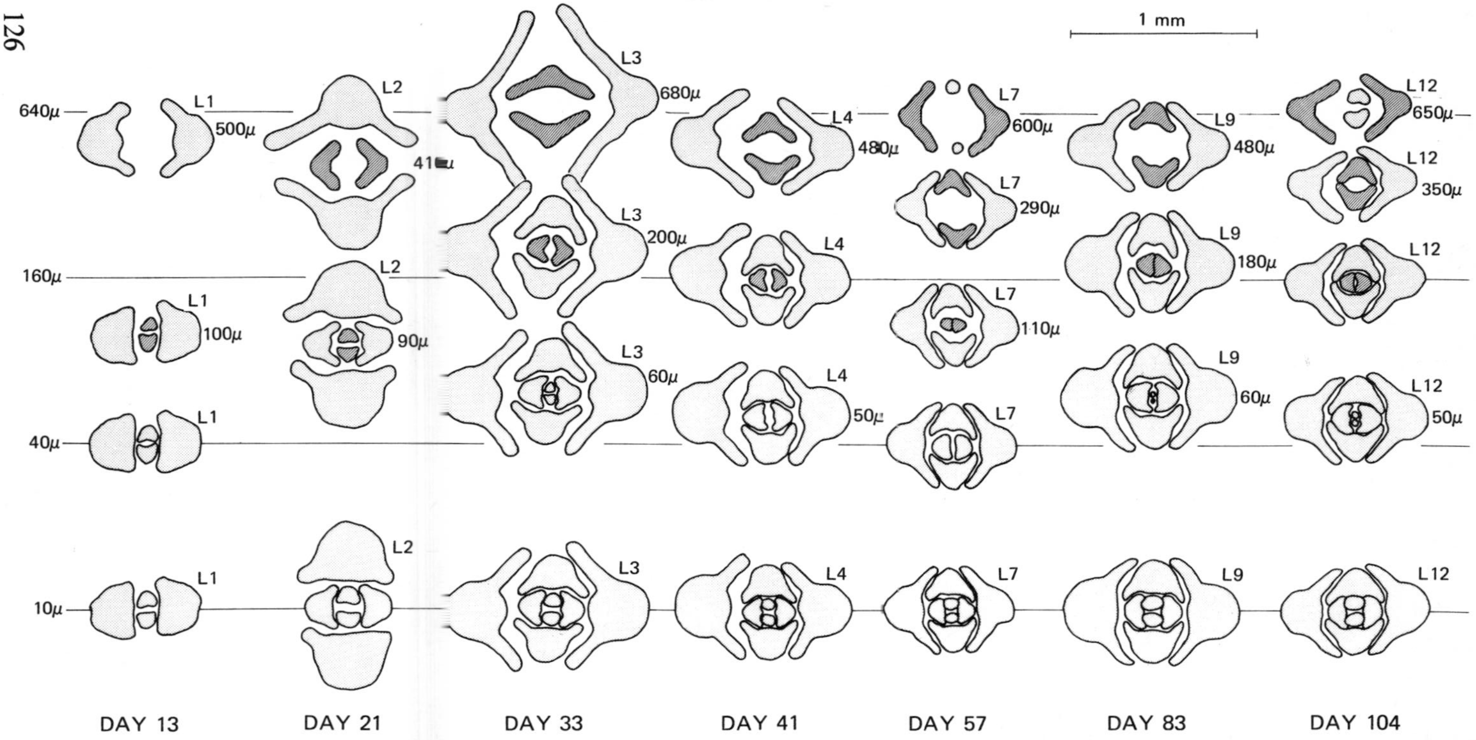

Fig. 4.6.9. Transverse section outlines for seven stages of seedling development in *E. bicostata*. The lowest row represents the cut surface of the first section which includes any part of the residual apex. The height of the other sections above the apices is usually given beside them, and may be gauged by their placement on or between the logarithmically spaced lines. Heavy stippling indicated that the leaf pair in question is sectioned half way between tip and half-junction (e.g. L3 on day 21 had a 0.8 mm blade).

Eucalyptus bicostata

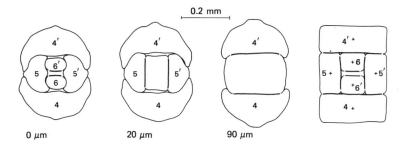

Fig. 4.6.10. The development of a transverse projection for a 'juvenile' apex of *E. bicostata*. Distances are in μm from the apex.

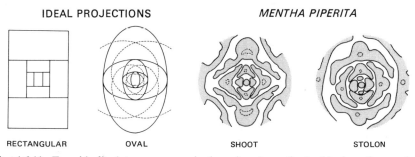

Fig. 4.6.11. Two idealised transverse projections based on the 'golden' section, and transverse section outlines for shoot and stolon apices of *Mentha piperita* L. These are based on Figs. 1 and 6 of Howe and Steward (1962).

inner and outer margins of primordium projections, as shown in the final diagram. Their distances apart across the centre, being twice their radii, can be used directly for calculating the plastochrone ratio, allowance being made for the fact that there are two leaves for each node. Thus the ratios, x, for the distances between centres for 4–5 and 5–6 are 1.531 and 1.992 respectively. The plastochrone ratio, r, is then given by $\log r = (\log x/2)$, (i.e. $r = \sqrt{x}$), and it is 1.237 and 1.411 for the two cases. That they differ by so much suggests that only that based on the two leaf pairs closest to the apex are acceptable. To test this possibility an analysis was made of values derived for successive stages of seedling growth in our two species. Mean values of r, with standard errors are as follows:

	E. bicostata	E. grandis
Based on:		
Youngest pairs	1.45 ± 0.026	1.53 ± 0.030
Next pairs	1.35 ± 0.044	1.35 ± 0.036
n (no. of observations)	5	6

127

Eucalyptus bicostata

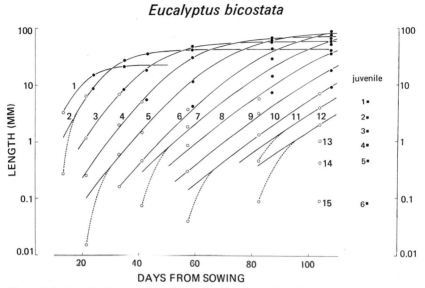

Fig. 4.6.12. Length–time relations for successive pairs of seedling leaves of *E. bicostata*. The column of values to the right of the figure are for the 'juvenile' axis of Fig. 4.6.8.

Appropriate statistical tests show the plastochrone ratio to be greater for the youngest pairs in each case. Thus for *Eucalyptus*, and possibly other decussate systems, it may not be safe to go beyond the first two leaf pairs to calculate plastochrone ratios. That this is so is made possible by precocious elongation of these axes, for this permits a rapid falling off in the radial relative growth rate of the attachment areas with age.

By contrast the two ideal transverse projections of Fig. 4.6.9 give the same plastochrone ratio throughout. In each case the ratio of successive radii is 1.618..., and $r = \sqrt{1.618...} = 1.272$. Similarly the ratios for successive radii are remarkably constant within the shoot and stolon diagrams for *Mentha piperita* L. These yield values of 1.324 and 1.219 respectively for r. It is realised, of course, that it is not usual to estimate r from single transverse sections, but longitudinal sections published in the same place (Howe and Steward, 1962) suggest that radial distances would have been much the same in properly prepared transverse projections of these apices.

We may now summarize these results in terms of the phyllotaxis index. For the seedling eucalypts the values are: *E. bicostata*, 2.27, *E. grandis*, 2.14; for older axes of *E. bicostata*: 'juvenile', 2.02, 'mature',

128

2.05; and for *Mentha piperita*: shoot, 2.57, stolon, 2.93. The value for the ideal transverse projections of Fig. 4.6.11 is 2.73, right between those for *Mentha*. However, all these values are considerably in excess of that for an orthogonal 2:2 system, which is 1.497, and would require the very large ratio of 4.81 for the successive radii of the leaf pairs. It is doubtful if such systems exist in nature. The four parastichy spirals of the ideal system (Fig. 4.6.9) actually intersect at about 143°, which could have been predicted by suitable interpolation in Fig. 3.7. These two ideal systems also illustrate the point made in Chapter 3 that numerically identical systems may have very different primordium shapes.

A final point concerning the development of phyllotaxis in *Eucalyptus* is that one can learn something of the rate of reconstitution of the bare apex from simple measurements of the transverse projection. Thus in Fig. 4.6.10 the gap between the fourth pair of primordia is barely twice that between the fifth pair, but the gap between the fifth pair is six times that between the sixth pair. These differences imply radial relative growth rates which are highest in the centre and decrease rapidly as the leaf pairs are initiated. From the formula $R_r = \log_e r/t_p$, where r is the plastochrone ratio and t_p the plastochrone, it may be shown that R_r can be as high as 0.18 per day for the bare apex at the beginning of the plastochrone when it is only 0.07 per day when based on the plastochrone ratio for the first two pairs of leaves. The implications of this will be considered further in Chapter 8.

Leaf growth

An array of length–time curves for successive leaf pairs is presented for *E. bicostata* in Fig. 4.6.12. This and the next two figures for volume–time relations lack the precision of most of the leaf growth data presented elsewhere because of the variability of the plant material. Median plants were used for the volume and early leaf length studies, but the remainder had to be used for dry weights and later mean lengths. The concept of plastochrone index developed by Erickson and Michelini (1957) was used to link the data together.

The first point to notice in Fig. 4.6.12 is that a number of the early values for specific leaf pairs are very low indeed, being far below any reasonable extrapolation from the later time courses for these leaf pairs. It is suggested that this phenomenon is simply the numerical expression of what has already been deduced from the three-dimensional drawings –

Eucalyptus bicostata

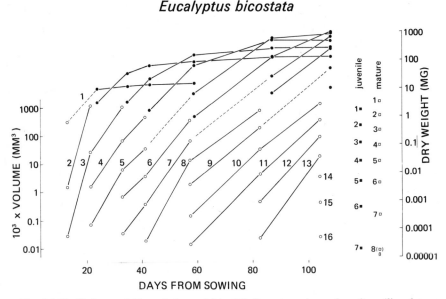

Fig. 4.6.13. Volumes (O) and dry weights (●) for successive pairs of seedling leaves of *E. bicostata*, plotted on equivalent logarithmic scales. The dry weight scale is extrapolated downwards to provide estimates based on assumed constancy of weight per unit volume. Volumes for the 'juvenile' (■) and mature (□) axes of Fig. 4.6.8 are shown to the right.

that the youngest pair of primordia grows freely until it is subject to the constraint of the preceding pair. This is why the logarithmic interval for length decreases rapidly with age within any one axis, and this pattern continues into the so-called 'juvenile' condition of older stems.

The volume–time relations of Figs. 4.6.13 and 4.6.14, support and supplement the length data. They show more clearly the differences in growth pattern for successive leaf pairs – differences which are consistent with the progressive increases in tightness of packing which have already been noted. The data are not good enough to justify the development of relative growth rate patterns such as those of Figs. 4.1.18, 4.2.10 or 4.5.9, but a few values for *E. bicostata* are of some interest. Early R_v values for leaf pairs 3, 7 and 11 are 0.821, 0.312 and 0.229 per day respectively. These older leaf pairs then settle down to a low rate of exponential growth for up to five weeks before maturation processes set in and reduce the rate to zero.

130

Eucalyptus grandis

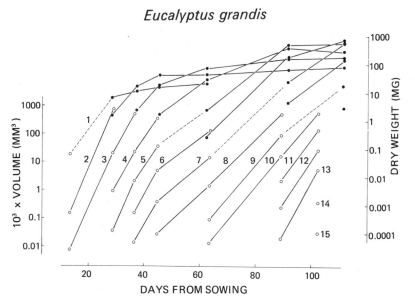

Fig. 4.6.14. Volumes (○) and dry weights (●) for successive pairs of seedling leaves of *E. grandis* plotted as in Fig. 4.6.13.

4.7 WHEAT, *Triticum aestivum* L.

Shoot-apical system

Because of the economic importance of cereals and grasses, studies of morphogenesis and histogenesis in gramineous plants have been quite extensive. The reader is referred to Sharman (1945) for a thorough and early account of leaf and bud initiation, and to Barnard (1964) for a general account of form and structure. Quantitative studies of vegetative growth have been made by the author and his colleagues (Williams, 1960; Williams and Rijven, 1965; Williams and Williams, 1968).

Figs. 4.7.1 and 4.7.2 provide perspective views of an early- and a late-stage, vegetative apex of wheat. Each leaf promordium arises well down on the flank of the apex. The bulge of tissue quickly becomes a ledge and spreads laterally to encircle the axis. This encirclement occurs within the limiting confines of preceding leaf primordia, and continuing growth is directed upwards, or parallel with the axis. These events produce first a collar and then a cowl, or hood-shaped structure which eventually overtops the apex, enclosing it and all subsequently-formed leaf primordia. Primordia at different stages in this process can be seen in the two text figures, and later ones in Fig. 4.7.13. Two features of the

131

Triticum aestivum

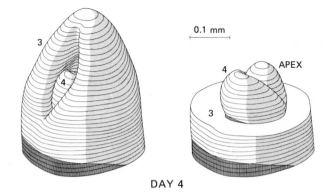

Fig. 4.7.1. Three-dimensional reconstruction of a 4-day apex of wheat. On the right, leaf 3 has been dissected away. Contour lines are 10 μm apart.

Triticum aestivum

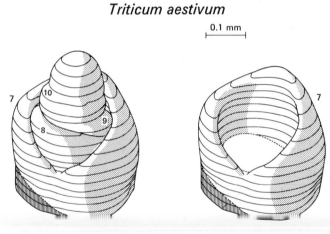

Fig. 4.7.2. Three-dimensional reconstructions of a 22-day vegetative apex of wheat. The dissection gives some idea of the tightness of packing of the primordia. Contour lines are 20 μm apart.

system call for special comment: first, that newly initiated primordia appear not to have any developmental space into which they may freely grow, even for a short time and, secondly, that the layered packing of the primordia is for a time so complete that it is difficult to visualize growth as being determined other than in terms of the properties of the system as a whole. However, the special point of Fig. 4.7.13 is to show that each leaf in turn is released from these constraints, and in such a

132

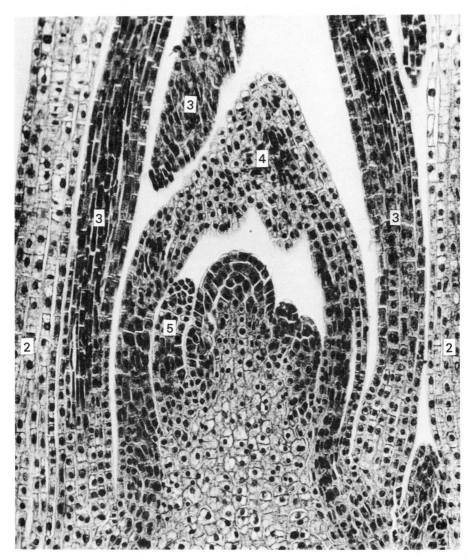

Fig. 4.7.3. Longitudinal section of an 8-day wheat seedling, showing the two-layered tunica of the growing point. The overtopping leaf is L4. (Williams, 1960, Plate 3.) (× 190.)

manner as to permit it to give expression to its latent potential for growth. This rather colourful interpretation will be shown to be consistent with the growth rate patterns within shoot apices of wheat.

That initiation and early growth of tiller buds in wheat also take place under conditions of considerable constraint is indicated by the drawings

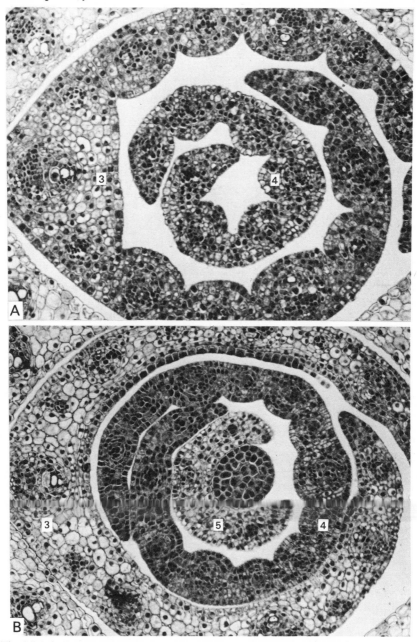

Fig. 4.7.4. A. Transverse section of an 11-day wheat seedling. The inner leaf is L4 section 87 of the same apex depicted in Fig. A.4. Most of L3 is also visible and is in an advanced state of vascular differentiation. B. Transverse section of the same seedling (section 39 of Fig. A.4). At the centre is the apex (heavily stained), L5 (lightly stained), and then L4 (heavily stained). (Williams, 1960, Plate 2.) (× 143.)

Triticum aestivum

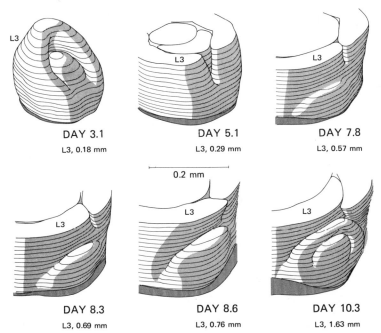

DAY 3.1
L3, 0.18 mm

DAY 5.1
L3, 0.29 mm

DAY 7.8
L3, 0.57 mm

0.2 mm

DAY 8.3
L3, 0.69 mm

DAY 8.6
L3, 0.76 mm

DAY 10.3
L3, 1.63 mm

Fig. 4.7.5. Three-dimensional drawings to illustrate the initiation and early growth of the tiller bud in wheat. The age equivalents of the stages assume the length growth of L3 to be exponential. Contour lines are 10 μm apart.

of Fig. 4.7.5. It should be remembered that leaf 2 or its sheath is tightly wrapped around leaf 3 and the bud in all the stages shown. Details of tiller bud initiation and the curious asymmetry of the bud-axis relation (see Fig. 4.7.6) are described by Williams, Sharman and Langer (1975) and the growth and growth rates of tillers as affected by light intensity and nitrogen nutrition are described by Williams and Langer (1975). This story is believed to be so important for the thesis that physical constraint is a significant determinant of morphogenesis and of growth rates of meristematic organs that it will be taken up again in Chapter 7. At this point it is sufficient to state that tiller buds appear initially to be moulded by, and determined as to rate of growth by the growth of the internode or, more precisely, of the cavity produced by the mutual growth of the bud, the internode, and the axillant leaf sheath. Whether or not a given bud will ever escape from the cavity and become an independent shoot system would seem to depend on whether its potential for growth can match the constraints of its physical surroundings. But more of all this anon.

Triticum aestivum

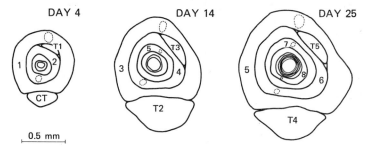

Fig. 4.7.6. Transverse projections of 4-, 14- and 25-day apices of wheat: T, tiller.

Phyllotaxis

In the previous chapter, *Dianella* sp. and *Triticum aestivum* were both given as examples of distichous phyllotaxis (Fig. 3.2E and F). From their transverse projections it is evident that that for *Dianella* is the simpler type and easy to measure for phyllotactic purposes. Indeed its projection in that Figure yields a plastochrone ratio of 1.73, which converts to a phyllotaxis index of 1.87. In one place Richards (1951) points out that orthogonality in the transverse plane for a distichous system (1:1 in his terminology) requires a plastochrone ratio of 23.14 and a phyllotaxis index of 0.057. Such conditions presumably never occur at plant apices. Interpolation within Fig. 3.7 shows that our *Dianella* apex would have logarithmic parastichy spirals which intersect at about 160°.

Transverse projections for three stages in the development of the wheat apex are given in Fig. 4.7.6. They were prepared from material used by Williams and Williams (1968) in their study of the effects of day length and light-energy level on many aspects of growth in the wheat plant. Further examination of that material yields a fairly complete statement of change in phyllotaxis, and this is summarized in Table 4.7.1. The essentials of the treatments used are set out diagrammatically:

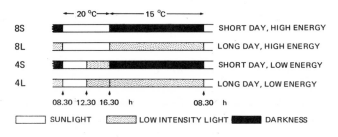

Table 4.7.1. *Plastochrone ratio and phyllotaxis index in wheat*

Leaf interval	Plastochrone ratio (r) Treatment				Mean	Phyllotaxis index (P.I.)
	8S	8L	4S	4L		
1–2	1.85	1.80	1.66	1.70	*1.77*	1.83
2–3	1.76	1.68	1.78	1.75	*1.75*	1.85
3–4	1.71	1.55	1.57	1.57	*1.61*	2.02
4–5	1.68	1.51	1.59	1.40	*1.56*	2.09
5–6	1.44	1.45	1.47	1.43	*1.45*	2.27
6–7	1.44	1.42	1.41	1.47	*1.44*	2.29
7–8	1.35	1.35	1.32	1.30	*1.33*	2.55
8–9	1.34	1.30	1.33	1.17	*1.31*	2.61
9–10	1.12	1.21	1.15	1.19	*1.17*	3.17
10–11	1.22	1.13	1.11	—	*1.16*	3.23
11–12	1.18	1.13	1.25	—	*1.19*	3.06
12–13	1.16	—	1.09	—	*1.13*	3.43

The double rules separate plastochrone ratios before and after the flag leaf.

Although the presence of large tiller buds distorts the projections and increases the errors of estimation, there are no consistent trends with time in plastochrone ratios based on a given pair of leaves. One is thus not limited to values obtained close to the apex as was the case for *Eucalyptus*. For the most part, values within the body of Table 4.7.1 are means of four or five values from stems of increasing age. Nor are there any consistent effects of the four treatments; even the foliar members of the inflorescence in the long-day treatments seem to have the same values as the corresponding foliage leaves of the short-day plants. In short, there are steady falls in the ratio with all treatments, and this is best represented by the column of general means. The corresponding rise in phyllotaxis index is also set out in Table 4.7.1.

However, the application of Richards's procedures of phyllotaxis measurement calls for some caution because of the curious shape of the apex. Figures 4.7.11 and 4.7.12 show that the apical cone progressively changes shape. The apical dome grows further and further away from the site of initiation of the most recent leaf primordium and the apex changes from a rather flat cone to an acute one. Indeed the apex proper eventually becomes a cylinder terminated by a dome. This means that transverse projections, though acceptable for early stages, are subject to progressive distortion. This contributes to the rapid fall of plastochrone ratios towards unity, and gives abnormally high values for the apex–primordium area ratio. These facts underlie the aberrant values

137

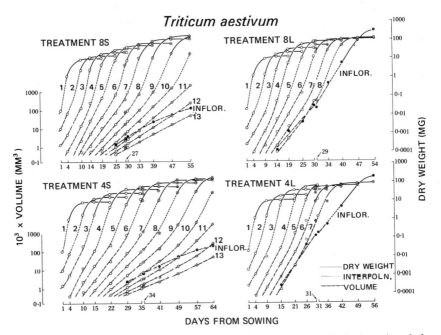

Fig. 4.7.7. Volume and dry weight changes for successive leaves (1, 2, 3, etc.) and the inflorescence for wheat plants subjected to two light-energy levels and two day-length treatments. The first dry weight values coincide with the times of emergence of the leaves. (Williams and Williams, 1968, Fig. 11.)

Table 4.7.2. *Relative length growth rates R_a within the apical cone* (day^{-1})

Tip of dome to:	Time interval (plastochrones*)				Mean
	1–2	2–3	3–4	4–5	
P_n**	0.232	0.182	0.130	0.165	0.177
P_{n-1}**	0.173	0.136	0.109	0.133	0.128
P_{n-2}**	0.147	0.111	0.089	0.113	0.115
	L.S.D. ($P < 0.05$) = 0.0151.				

* Based on leaf emergence.
** P_n is the half-junction of the most recently initiated primordium at the beginning of each interval. P_{n-1} and P_{n-2} are the second and third such junctions. Compare the diagram of Fig. 4.1.1B.

for wheat in the relation of Fig. 3.8. They are for later stages in development, and the area ratios are relatively more disturbed by the distortion than are the phyllotaxis indices.

138

Triticum aestivum

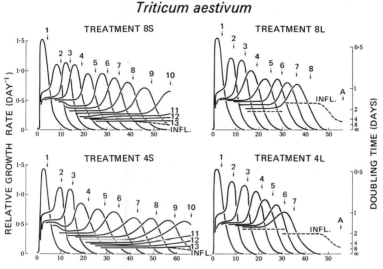

Fig. 4.7.8. Relative growth rates for successive leaves (1, 2, 3, etc.) and the inflorescence of the primary shoot of wheat plants subjected to two light-energy levels and two day-length treatments. The arrows mark the times of emergence of the leaves, and A indicates the times of anthesis in the long-day treatments. (Williams and Williams, 1968, Fig. 12.)

Axial growth

It has been suggested for flax that the relative rate of extension growth is higher in the apical dome than it is in the sub-apical region, at least during early vegetative growth. It is of interest, therefore, to see if this is true for wheat, where these two regions are more easily distinguished. To that end our data for the two short-day treatments were subjected to the same analysis as that indicated for flax in Fig. 4.1.13 and Table 4.1.4. The results are set out in Table 4.7.2, and this time it is possible to test the significance of differences between the three methods of estimating the rates. This was because neither treatment nor its interaction with time had a significant effect, and the second order interaction could be used to test the interaction between method and time. The values in Table 4.7.2 are means for the two short-day treatments of Williams and Williams (1968).

The result is quite clear cut, for the rates based on P_n are always greater than those based on P_{n-1} and those on P_{n-1} are greater than those on P_{n-2}. That there are also effects of time and an interaction with method is of no special interest here. The result can only mean that,

139

Triticum aestivum

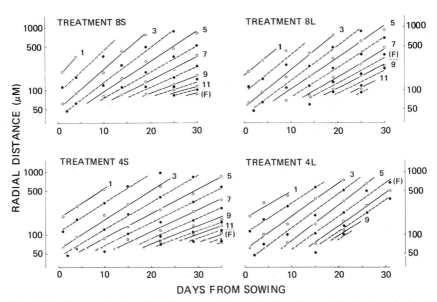

Fig. 4.7.9. Changes in the radial distances of the mid-veins of successive leaves of wheat plants subjected to two light-energy levels and two day-length treatments. F marks the flag leaf in each case.

for all stages examined, there is a strong gradient in R_a from a high value above P_n to a low value at P_{n-2}. Indeed, a direct estimate for the internodes between P_n and P_{n-2} gave a mean value of only 0.045. This zone is, of course, one of intense morphogenetic activity – mainly lateral activity connected with the establishment of new leaf primordia (see Fig. 4.7.2). A renewal of extension growth comes with the establishment of an intercalary meristem and subsequent internode growth.

Leaf growth

Fig. 4.7.7 is from Williams and Williams (1968) and gives arrays of growth curves for their four treatments in terms of volume and dry weight. That the individual curves are sometimes built on very few experimental values should not be a matter for concern, because their general form is well established by earlier work (see Williams, 1960, p. 416; and Williams, 1966a, p. 956). That general form, which covers about five logarithmic cycles in size, is repeated in Fig. 4.7.14. The essentials are a period of near-exponential growth followed by an increase to a maximum rate just prior to leaf emergence, and then a

140

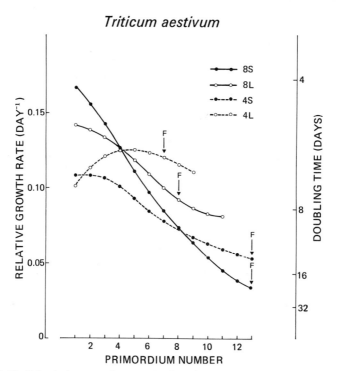

Fig. 4.7.10. Radial relative growth rates, R_r, based on the slopes of the arrays of lines in Fig. 4.7.9. F indicates the flag leaf for each treatment. The points on the curves are interpolations, not experimental values (see text).

rather slow decline in rate as maturation proceeds. It is perhaps best to try to grasp the full pattern of growth at the shoot apex, and the effects of treatment thereon by examining Figs. 4.7.7 and 4.7.8 concurrently. These present the data on logarithmic scales and as relative growth rates.

The period of exponential growth for the first three leaves may be presumed to have commenced during embryo growth, being interrupted by seed dormancy. Thereafter the duration of exponential growth varies from a few days in early leaves to as much as four or five weeks in leaves 12 and 13 of the short-day plants. The exponents fall only slowly with leaf number in the long day plants, more rapidly with 8S and most of all with 4S. In spite of these variations in pattern, the sizes of the leaves at the end of their exponential phases are little affected by treatment, though they increase with leaf number. At the high-energy level, the number of foliage leaves on the primary shoot is reduced from 13 to 8, and at the low-energy level from 13 to 7 by the long-day treatment.

Following the early exponential phase of growth, the relative growth

Triticum aestivum

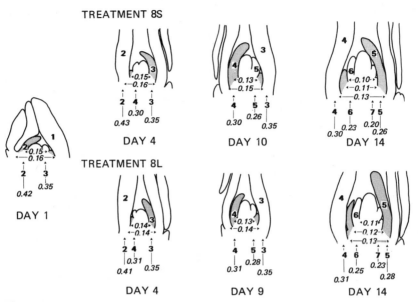

Fig. 4.7.11. Outlines of median longitudinal sections for four stages of early seedling growth as affected by short- and long-day treatments. Leaf numbers are given in bold type, and length and radial relative growth rates are in italics. The stippling indicates the primordium which has most recently overtopped the apex.

rates of the leaf primordia rise to maxima whose values are approximately twice those for the exponential phase. The maxima occur two or three days before leaf emergence, and the rates then fall to zero. The maxima decrease with increasing leaf number, and the duration of successive phases of the growth rate curves increases considerably.

The studies of phyllotaxis already reported yielded extensive data on the radial distances of leaf primordia at their junctions with the axis. These are presented in Fig. 4.7.9 and give useful information on radial growth rates near the apex (cf. Fig. 4.1.11 for flax). They are well fitted by families of linear regressions using principles laid down in the appendix. Doubtless these relations would soon depart from linearity if extended further. Radial relative growth rates are given by the *b* values of the regressions, and these are presented as a function of primordium number in Fig. 4.7.10. Treatment interactions are strong and rather complex. However, comparison of Figs. 4.7.7 and 4.7.9 shows that early volume and radial growth rates are highly correlated. Williams and Williams (1968) drew attention to the remarkable agreement between the initial relative growth rates for the inflorescences and

Triticum aestivum

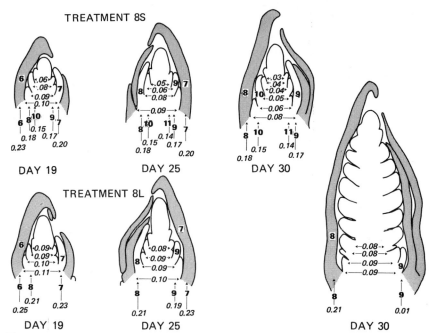

Fig. 4.7.12. Outlines of median longitudinal sections for three later stages of growth as affected by short- and long-day treatments. Details as in Fig. 4.7.11.

those for the youngest leaves (the flag leaves) with which they are associated. It is suggested that a thorough understanding of the transition from the vegetative to the reproductive condition could profit by a close examination of these growth rate patterns.

Figs. 4.7.11 and 4.7.12 together attempt to integrate the information for the contrasting short- and long-day treatments (at the high energy level) with morphological change within these treatments.

Unfortunately these diagrams do not include axial growth rates such as those given in Table 4.7.2 for the short-day treatments. The material proved to be too variable to obtain satisfactory values for the long-day treatments. The patterns are almost identical to day 14, but radial growth rates remain high from day 19 onwards in long days, and continue to fall in short days. That axial growth rates are also very low in the sub-apical region of short-day plants (Table 4.7.2) could be a valuable key to an understanding of the continuance of foliar dominance. I realize that such a notion might not readily gain acceptance, but it is clear that a differential involving physical constraint could well determine whether a potential flag-leaf primordium would be followed by a normal foliage leaf (short days) or by a diminutive foliar ridge (long days). The leaf length story for these contrasting transitions is set out in Tables 6 and 7 of their paper by Williams and Williams (1968).

We can now return to the generalized curves for leaf growth and relative growth rate of Fig. 4.7.14. From time to time over a period of nearly fifteen years, I and my colleagues have sought explanations for the forms of these curves. Why is growth exponential at first, sometimes for as long as four weeks? And why, when the leaf attains a size which is specific for that leaf, does the rate rise to a value which is approximately twice that during the exponential phase? Quite early it was noticed that the first protophloem elements were differentiated towards the end of the exponential phase in all leaves, and irrespective of treatments (Fig. 4.7.4). This correlation seemed to suggest that the subsequent increases in rate were due to a more efficient supply of the necessary substrates for growth. However, later work showed that the same event marked the *entry* of the tobacco leaf into a phase of exponential growth (Hannam, 1968), and, in subterranean clover, the event occurs in the centre of a very extended period of exponential leaf growth (Williams and Bouma, 1970). A second correlation noted was that, with the rise in R_v, there was an increase in the stainability of the cytoplasm of the primordium. Williams and Rijven (1965) have shown the increase to be due to an increase in the concentration of RNA (Figs. 4.7.3 and 4.7.4) and that there was also an increase in the rate of protein synthesis. However, they also concluded that these were effects rather than causes of the increase in R_v. Growth substances were also invoked at one stage in the argument: clearly they must be involved in these sequences of events, but there seemed no good reason to assume them to be causally involved.

It was not until the parallel study of subterranean clover drew attention to the fact that successive leaf primordia of that plant followed each other in a developmental tunnel that it occurred to us that events at higher rather than these lower levels of organization might provide the key to understanding. Was it possible that physical constraints implicit in shoot apices as dynamic and closely integrated systems would supply that key? With this in mind we can look again at Fig. 4.7.13.

On day 14, leaf 6 is tightly packed inside leaf 5 and the enlarged inset to the figure shows that it is also very tightly wrapped around leaf 7 and the rest of the apex. It is almost self-evident that leaf 6 can only grow at the rate permitted by leaves 5 and 7 and the apex. On day 19, the situation of leaf 6 is not much improved, in this respect, for its lower

144

Triticum aestivum

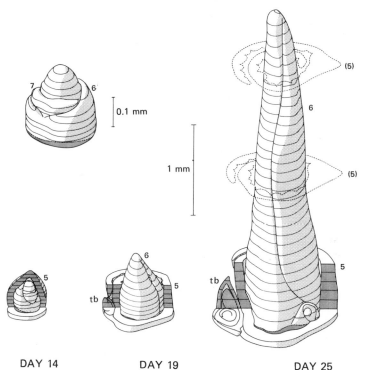

0.1 mm

1 mm

DAY 14   DAY 19   DAY 25

Fig. 4.7.13. Three-dimensional drawings illustrating the relations of leaf 6 to other elements of the shoot apex on days 14, 19 and 25. The inset is for leaf 6 and the apex on day 14, drawn to a larger scale: tb, tiller bud.

half is still tightly clasped by leaf 5 and, though this is not shown in the drawing, its abaxial surface is in contact almost to its tip. By day 25, however, the situation is very different. Contact with leaf 5 is limited to less than 0.5 mm at its base. Therefore the greater part of the abaxial surface of leaf 6 is now free of leaf 5, as is indicated by the 'shadow' transverse sections of Fig. 4.7.13. One has only to assume that the growth rate of leaf 6 prior to day 25 is less than its potential rate to realize that the rapid change in the configuration of the space available for development implicit in the drawings will permit leaf 6 to grow more rapidly. The juxtaposition of miniatures of the same drawings with the growth and relative rate curves for leaf 6 in Fig. 4.7.14 will possibly convey the point even more forcefully. Why it is that the rate of growth while the leaf is in a state of dynamic constraint should approximate to the exponential is a difficult question, and will be taken up again in

145

Triticum aestivum

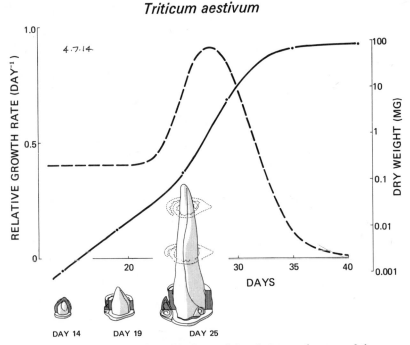

Fig. 4.7.14. Schematic presentation of leaf growth in relation to the state of the apex as depicted in Fig. 4.7.13.

the sequel. That the exponent should decrease with leaf number and at varying rates according to treatment is also quite intriguing, and suggests systematic variation in other properties of the system. For instance, for comparable stages of length growth, leaf primordia are thinner with increasing leaf number (compare leaf 7 in Fig. 4.7.2 with leaf 3 in Fig. 4.7.5), so there will be gradients in surface–volume relations, and these could vary with treatment. It might be as well to end this section before one is tempted to indulge in wilder flights of fancy.

4.8 FIG, *Ficus elastica* Roxb. ex Hornem

Shoot-apical system

When this apex was selected for investigation the author had already noticed for clover and wheat that exponential growth went with close packing of primordia within apical systems, and that it was not an obvious feature in a relatively lax system such as that of flax. It seemed logical, therefore, to select another system in which close packing was

Fig. 4.8.1. A. Shoot apex of *Ficus elastica* with the youngest, fully-emerged leaf in the background. The stipule of this leaf completely covers all younger parts. (×0.5.) B. The same apex a few days later, when the stipule, now reddish brown and senescent, will soon be shed. The next leaf is in process of unrolling. (×0.5.)

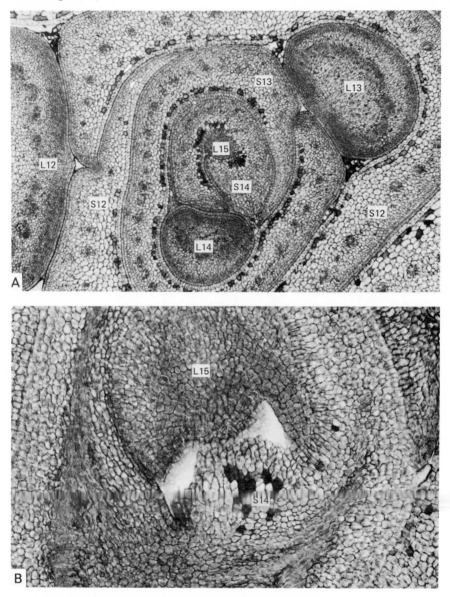

Fig. 4.8.2. A. Transverse section of an apex of *Ficus elastica* taken 0.18 mm above the residual apex. Leaves and stipules marked L and S respectively. (×81.) B. Transverse section through the residual apex of the same axis. The interpretation of these tightly packed structures can be quite difficult in such sections. (×220.) The line drawing of Fig. 4.8.3 is based on this section.

Ficus elastica

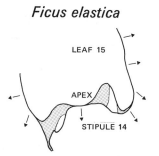

Fig. 4.8.3. Explanatory diagram for the centre of Fig. 4.8.2B. Full lines are edges or junctions on the upper surface of the section. Stippling indicates free epidermal surfaces which pass out of focus in the photo-micrograph. Arrows indicate direction of slope of epidermal surfaces actually in contact. Note especially that stipule 14 is actually in contact with the residue of the apical dome at the centre of the picture.

an obvious feature. *Ficus elastica* filled this requirement extremely well because the specimens examined (Fig. 4.8.1) had eight or nine leaf primordia packed inside a very large conical structure which is made up of the fused stipules of the youngest fully-emerged leaf of the axis. Since it was not practicable to produce hundreds of such apices in the available controlled environment space, we managed with only eight. This was possible because our clonal material produced leaves of very uniform size at approximately seven-day intervals, and we were able to use the concept of age equivalence to great advantage. Since the leaf and stipule were so easy to separate it was decided to measure and weigh them separately.

The extraordinarily tight and very beautiful packing of the successive members of the apex is perhaps best illustrated by the transverse section of Fig. 4.8.2A and by the family of transverse-section outlines of Fig. 4.8.4. The three-dimensional drawings of Fig. 4.8.5 also show the relations within the inner part of the bud. The time scale implied on the drawings of this section usually refers to the largest leaf of that drawing, and is an estimate of the time lapsed since that leaf was initiated. Thus in Fig. 4.8.5A the largest leaf and its stipule (partly removed in the drawings) is estimated as 34 days old. The second leaf and its stipule can be taken to represent a 27-day leaf, the third (at B) as a 20-day leaf, and so on. The stipule is quite an extraordinary structure, for its two members are fused most of the way up on the side away from the leaf; and separate, though tightly overlapped, along their entire inner margins. It was not until we sought to interpret curious sequences of sections near the tips of the stipular structures (see Fig.

149

Ficus elastica

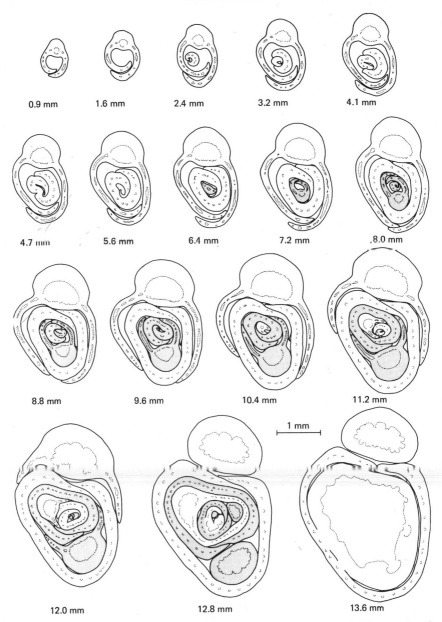

Fig. 4.8.4. Outline drawings of transverse sections of another and somewhat larger apex of *Ficus elastica*, mainly to illustrate the beautiful pattern of packing of the leaves and their stipules. The second and fourth leaves are distinguished by stippling.

4.8.6) that we realized that they represented the tips of fused stipules. At day 34 these are very unequal in size and the larger almost encloses the smaller. The genesis of form in the stipular structure is dealt with below. Another feature of the system which is worthy of note at this point is that the laminae of the leaf primordia are relatively late comers to the scene, and make up an increasing proportion of the substance of the developing leaf. Just before emergence the lamina is wrapped several times around itself, one wing over the other.

The three-dimensional drawings of Figs. 4.8.7, 4.8.8 and 4.8.9 illustrate the genesis of form in the leaf and stipule. The earlier ones were rather difficult to prepare simply because the sections themselves were difficult to interpret, as will be seen from Fig. 4.8.2B and the line drawing of Fig. 4.8.3. When surfaces in contact are cut transversely or nearly so, the double lines of epidermal cells make identification of the boundaries a simple matter (Fig. 4.8.2A), but when they are cut at 45° or less it is necessary to work with a high power of the microscope and to focus up and down.

The drawing for day 7 (Fig. 4.8.7) is believed to represent the end of the plastochrone which produced the large primordium pictured. The stipular arms almost encircle all that is left of the spherical surface of the apex proper, and they are already unequal in size. Half a day later (day 7.5) the discrepancy between the stipules is more marked, and the bulge of the next primordium has appeared (with an age equivalent of 0.5 of a day). Unfortunately there is a five-day gap in the story which leads to the left-hand drawing of Fig. 4.8.7, and to the re-orientation of the sloping residual apex through the Fibonacci angle of 137.5°. In seeking to understand the contortions involved in the genesis of form during these early stages one needs continually to remind oneself that the whole is enveloped in the stipule of the previous primordium. Even on day 7, there is direct contact between it and the residual apex, but *not* where the new primordium will arise.

By day 11 (see Fig. 4.8.8, but note the reduced scale) the leaf and stipules have both grown considerably. The leaf now leans inwards and fills much of the cavity below the previous set of stipules: its marginal meristems have also become active. The stipules now fill the greater part of the space under the leaf and the remaining space under the previous stipules: the smaller of the stipules is also beginning to be surrounded by the larger.

Two days later (Fig. 4.8.9) the same processes have continued, and the three drawings, with their progressive dissections, give a very clear

Ficus elastica

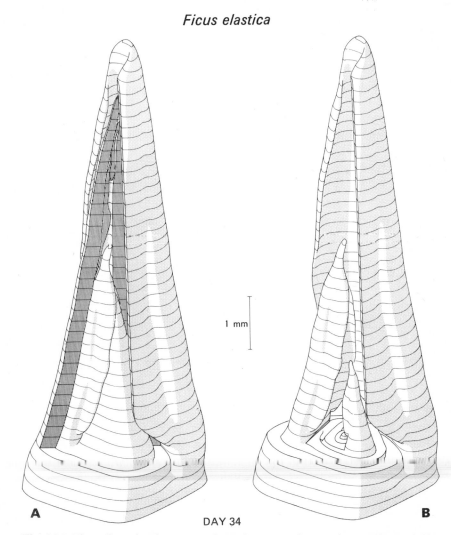

1 mm

A

B

DAY 34

Fig. 4.8.5. Three-dimensional reconstructions of an apex of *Ficus elastica*. The age is the estimated number of days since the initiation of the large outer leaf. In A the near half of the stipule of that leaf and portion of its leaf blade are dissected away to show the next leaf and its stipule. In B the first four stipules are removed to show the spatial relations of the leaves. The fourth leaf is hidden behind the third.

Ficus elastica

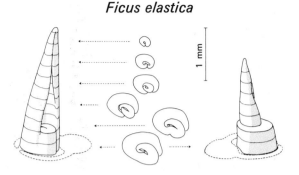

DAY 34

Fig. 4.8.6. Detail of the same apex as that represented in Fig. 4.8.5. It shows the relations of the unequal tips of the stipules. The orientation of the two main drawings is indicated by the 'shadow' outlines of the appropriate transverse section of the leaf.

idea of the relation between the stipule and the succeeding leaf primordium (Fig. 4.8.9C). That primordium has an age equivalent of six days, so is between those depicted for days 5.5 and 7.0 in Fig. 4.8.7. In the light of this account it seems reasonable to claim that the changing shape of the developmental space above the apex plays a significant role in the genesis of form both of the leaf primordium and its compound stipule. Whether this role can be described as dynamic physical constraint is a matter which calls for further thought.

Phyllotaxis

The apical cone of *Ficus* is so flat that it is fairly easy to produce a transverse projection such as that of Fig. 4.8.10. Since each primordium completely surrounds its successor, the only possible contact parastichy spiral is the genetic one. Church's terminology cannot cope with such a system, even though it displays the Fibonacci divergence angle of 137.5°. The plastochrone ratio is 1.692 and this converts to a phyllotaxis index of 1.91. The system is therefore rather close to a 2:3 orthogonal system in Richards's terminology.

Leaf growth

To prepare the growth curves of Fig. 4.8.11 it was first necessary to establish length–time relations for both leaf and stipule. These were

Ficus elastica

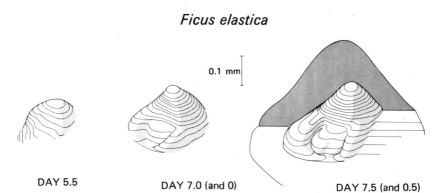

DAY 5.5   DAY 7.0 (and 0)   DAY 7.5 (and 0.5)

Fig. 4.8.7. Three-dimensional reconstructions of very early stages in the morphogenesis of the leaf primordium. The undefined boundaries on the near and right-hand sides of the primordia are in fact junction lines with the previous stipule. This stipule is in close contact with the primordium in the manner indicated for day 7.5.

Ficus elastica

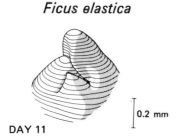

DAY 11

Fig. 4.8.8. A few days later, when the unequal stipules have covered the apex and fill the space between the associated leaf and the previous stipules.

Ficus elastica

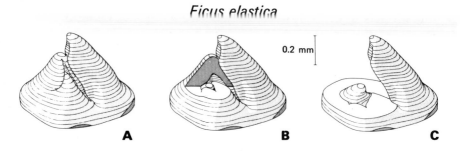

A   B   C

DAY 13

Fig. 4.8.9. Three-dimensional drawings illustrating the progressive dissection of a 13-day apex to show its relation to the next primordium and the residual apical surface.

154

Ficus elastica

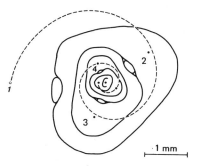

Fig. 4.8.10. Transverse projection and genetic spiral of the shoot apex.

based mainly on the three plants used for volume estimation and assumed a plastochrone interval of seven days. Their logarithmic plot
was slightly curved throughout the period of active growth and then
turned over to mature limits of 300 and 250 mm for leaf and stipule
respectively. All available volume and fresh weight values were then
plotted against their age equivalents. For example, a leaf primordium
between 1.91 and 2.22 mm in length and a stipule between 1.49 and
1.72 mm both have an age equivalent of 20 days.

The outcome in Fig. 4.8.11 shows that leaf growth is exponential for
more than 40 days, and covers in that time a size range of six logarithmic
cycles. The stipule has a smaller volume at first – presumably because
it starts later (Fig. 4.8.7) – but it grows faster, and actually exceeds the
parent leaf in volume for a short time. However, as the developmental
space into which it is growing fills up, the relative growth rate of the
stipule declines and settles down for about three weeks to an exponential
rate which is appreciably less than that for the leaf over the same period.
During their respective exponential phases of growth the doubling time
for the leaf is 2.42 days and that for the stipule is 3.15 days.

One naturally asks how it is possible for two dissimilar structures
growing in close contact to have different relative rates of growth. The
answer would seem to reside in their differences in form. The stipules
form hollow cones whose apical angles get progressively less with age.
Provided their relative growth in thickness is similar to that for radial
growth of the whole bud – and this seems to be so (Fig. 4.8.3) – it
follows that the spaces between the cones must grow relatively faster
than do the cones themselves. It is of little importance whether one
argues that the marginal leaf meristems passively fill or actively create

155

Ficus elastica

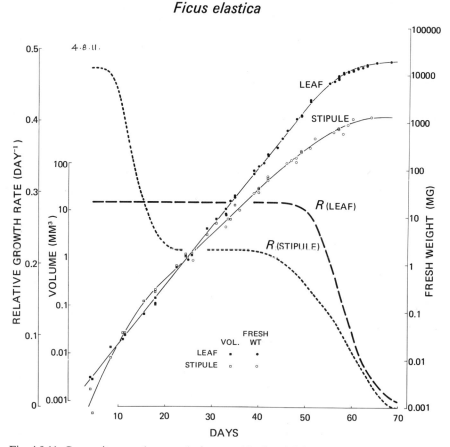

Fig. 4.8.11. Composite growth curves (volume and fresh weight) and relative growth rates for leaf and stipule separately. The fresh weight scale is extrapolated downwards to provide estimates for the early stages. The plants were grown in a controlled temperature environment of 27°/22 °C.

this extra space; the net result is a very neat and efficient solution to a developmental problem. The solution clearly requires the existence of sliding or gliding growth between lamina and stipule and lamina and lamina, but where successive stipules remain in contact growth could well remain synchronous.

Clearly the study of the growth of the terminal bud of *Ficus elastica* has yielded two very remarkable examples of exponential growth, and much food for thought on the possible relevance of constraint on growth rates and on the genesis of form in leaf and stipule.

156

4.9 YELLOW SERRADELLA, *Ornithopus compressus* L. *Dianella* sp. (*Liliaceae*)

NARROW-LEAF WATTLE, *Acacia mucronata* Willd. ex H. Wendl.

SUNFLOWER, *Helianthus annuus* L.

This final section of the chapter deals with four shoot-apical systems which were not examined in detail, but which illustrate points of some interest.

Serradella is representative of plants with a pinnate leaf structure, and seedlings were examined mainly to see how the pinnae are initiated. Fig. 4.9.1A shows a four-day apex in which the residual apical dome is a small saddle-shaped area between L2 and L3. Both of these primordia encroach almost to the summit of the dome and the initiation of the next primordium (L4) must await the further reconstitution of the bare apex by its growth upwards and away from L2. The system is distichous, or perhaps spirodistichous as in subterranean clover.

Figure 4.9.1B shows that, as one might expect, the pinnae are initiated along the two inner margins of the primordium. These same margins produce the marginal meristems of an entire leaf (cf. *Ficus* above). How it is that the pinnae are initiated at uniform intervals of 0.03 mm is an intriguing problem, as is the determination of the number of pinnae on a given leaf. Fig. 4.9.2 shows later steps in the genesis of the form of the pinnae, and Fig. 4.9.2A reminds us that the terminal pinna is a product of the main axis of the leaf. It bears the same relation to the rest of the pinnae as does the terminal spikelet to the lateral spikelets of a wheat ear.

A transverse projection for *Dianella* sp. (Liliaceae) was given in Fig. 3.2 as a fairly typical example of distichous phyllotaxis. This particular axis has been shown to have a phyllotaxis index of 1.87 (p. 136). The leaf is a curious combination of forms, for it is equitant at its base, as in *Iris* throughout, but opens out above to give a strap-shaped, dorsi-ventral leaf. Fig. 4.9.3A shows that the two halves of the blade of an early-stage primordium are folded together, and the whole is concealed in the deep but very narrow pocket in the base of the previous leaf primordium. Primordia 6, 7 and 8 and the apex are within L5 and are drawn separately to a larger scale in Fig. 4.9.3B. The eighth primordium is very small and out of sight on the far flank of the apex. These drawings leave little room for doubt that the genesis of form in this shoot-apical system is

157

Ornithopus compressus

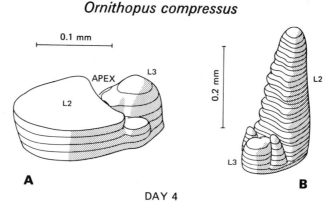

DAY 4

Fig. 4.9.1. Short apex of a 4-day seedling of serradella. A. Large scale three-dimensional drawing with L2 and its stipules removed, L3 intact. B. Another view of the same apex on a smaller scale, and with L2 and its stipules intact.

Ornithopus compressus

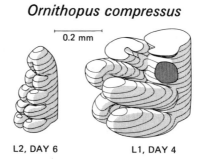

L2, DAY 6   L1, DAY 4

Fig. 4.9.2. Three-dimensional drawings of portions of leaves of serradella. Tip of L2 on day 6 and a few central pinnae of L1 on day 4. One pinna removed to show inner surfaces of other pinnae.

dominated by the changing form of the space available for development. Nor would it be particularly fanciful to predict that leaf growth is strictly exponential from a very early stage.

The preparation of the drawings of Fig. 4.9.4 was prompted by the desire to trace the origin of the rather beautiful stellate pattern of Fig. 4.9.5. This shows a transverse section near the apex of *Acacia mucronata*, one of the many Australian species of the genus whose leaves, at least in the mature plant, are really isobilateral phyllodes. The transverse section suggests a (5 + 8) system of phyllotaxis, with superficial resemblances to the diagram B in Fig. 3.3. The resemblance rests, of course, on the shapes of the primordia away from their points

158

Daniella sp.

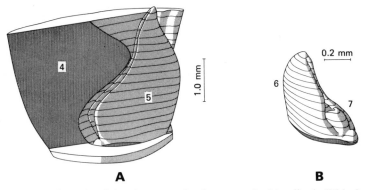

A   **B**

Fig. 4.9.3. Three-dimensional drawings near the shoot apex in *Dianella*. A. With the near half of the lower part of L4 removed to show the snug seating of L5 within its base. B. L6, L7 and the apex drawn to a larger scale. The bulge behind L7 is the bud in the axil of L5. The contour lines are 320 μm apart in A and 40 μm apart in B.

Acacia mucronata

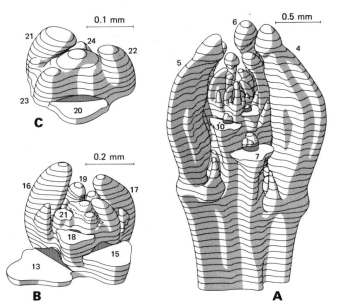

Fig. 4.9.4. Three-dimensional drawings of the apex of narrow-leaf wattle. A. General view with L7, L10 and L12 removed to show relationships within the apex. Contours 80 μm apart. B. Closer view with L13, L15, L18 and L21 removed. Contours 20 μm apart, except near the apex where a few 10 μm contours are shown. C. View of apical dome and initiating leaf primordia. Contours 10 μm apart.

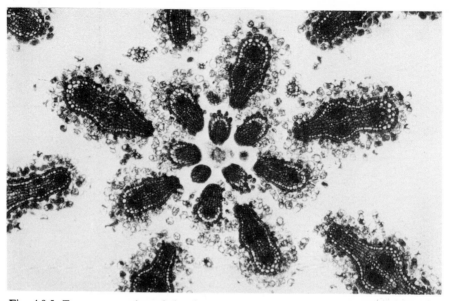

Fig. 4.9.5. Transverse section of the shoot apex of *Acacia mucronata* showing stellate arrangement of primordia around the apical dome. All but the youngest primordia are covered with short, clavate glandular hairs. (× 92.)

of attachment. Fig. 4.9.4 shows that the stubby stipules would make an inordinately large contribution to the transverse projection. One prepared from the present axis looks rather like that for tobacco (Fig. 4.2.6, day 31); it has a plastochrone ratio of 1.155 and hence a phyllotaxis index of 3.26. This is somewhat higher than that for a 3:5 orthogonal system, and the parastichy spirals would intersect at about 105°.

By comparison with most of the shoot apices discussed in this chapter, this apex seems curiously exposed and vulnerable. However, terminal meristems of this plant are only active when soil moisture is plentiful. It would be interesting to know if the glandular hairs of the young primordia have a role in protecting the apex.

The sunflower capitulum figured in Fig. 3.2 as an example of a high-order Fibonacci system, and the photo from which that diagram was prepared is reproduced as Fig. 4.9.6. There is small wonder that this capitulum should have become something of a classical object for the study of phyllotaxis. Not only is the spiral arrangement easy to demonstrate, but florets are already in a transverse plane. The corona of floret buds in Fig. 3.2C contains 238 buds and yielded 34 independent estimates of the plastochrone ratio. Its mean value for the corona is

160

Fig. 4.9.6. Capitulum of sunflower, *Helianthus annuus*, showing a pronounced (34 + 55) Fibonacci spiral system. (By courtesy of Mrs Joan Simpson.)

1.00223, and this yields a phyllotaxis index of 7.59. Even across this corona, however, there is almost certainly a gradient in the ratio, for the more obvious contact parastichies are $(34+55)$ in its outer part and $(21+34)$ in its inner part. And it can be safely predicted that towards the centre of the capitulum they would quickly change to $(13+21)$, $(8+13)$ and so on. This whole process is a reversal of what happens in the flax apex where, with a slowly-increasing, bare, apical surface, and a fairly constant primordium size at initiation, the system rises steadily through the lower members of the Fibonacci pairs. In the sunflower capitulum, on the other hand, the later stages of differentiation occur on a bare surface of diminishing diameter, and with primordia of constant initial size. These two systems demonstrate very clearly that spiral systems do not become more or less complex in a step-wise fashion, but change continuously with change in relevant parameters.

5 The dynamics of leaf growth

Much has already been said about leaf growth, because this is essential to the definition of the patterns of growth within shoot-apical systems. For two of those systems, however, Williams and Rijven (1965, 1970) have provided descriptions of growth and chemical change which relate to leaf growth as such.

Leaves are organs of limited growth arising laterally on the shoot apex, and they exhibit a steady progression from the meristematic condition of the primordium through to maturity and senescence. It is scarcely surprising, therefore, to find strong similarities in this progression even for such different plants as subterranean clover and wheat. Both studies happen to be on the fourth leaf of the primary shoot, and they are concerned mainly with changes in DNA, RNA, proteins and cell wall materials. The plants were grown in controlled environments, and the work achieved a high level of precision through sampling methods and chemical procedures which are fully described in the original papers. This chapter summarizes the main results.

5.1 SUBTERRANEAN CLOVER

The fourth leaf of subterranean clover was chosen because it is the first on the main axis to attain the characteristic form and size of the adult trifoliate leaf, and because its development could scarcely have been influenced by cotyledonary reserves. Fig. 5.1 illustrates the changing size and form of the fourth leaf from day 7, the earliest stage for which cell counts are available. The first chemical data relate to day 10, when the primordia are still only 0.66 mm long. The drawings of Fig. 5.1 include a thin disc of stem tissue, the disc of leaf attachment, but this was not in the samples. The petiole arises as a constriction below the three folioles on day 9 and quickly establishes the columnar form of days 11 and 13. There are no major changes of form after day 13, though the petiolules are established before the leaf emergence on or about day 16. Thereafter the petiole continues to elongate and the leaf blades spread as full maturity approaches. The differences in the time courses of length growth between blade and petiole are shown in the

Trifolium subterraneum

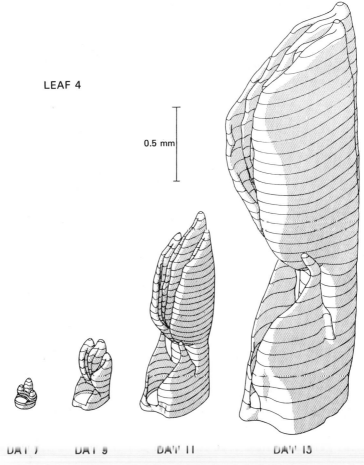

LEAF 4

0.5 mm

DAY 7   DAY 9   DAY 11   DAY 13

Fig. 5.1. Three-dimensional reconstructions of the fourth leaf of subterranean clover at 2-day intervals. The contours are 20 μm apart for day 7, 40 μm for days 9 and 11, and 80 μm for day 13. (Williams and Rijven, 1970, Fig. 2.)

inset to Fig. 5.2. Growth during days 19–25 is due almost entirely to petiole extension.

The length–time relation of Fig. 5.2 has a logarithmic scale and is a composite of the linear regression established in Fig. 4.5.6 of the last chapter, and a freehand curve through subsequent mean values at harvest. The size distribution histograms show that length variation relative to the sample mean increases to a maximum on day 14, where there is a range of 2.5 to 8.9 mm. There is even a suggestion of a bimodal

164

Trifolium subterraneum

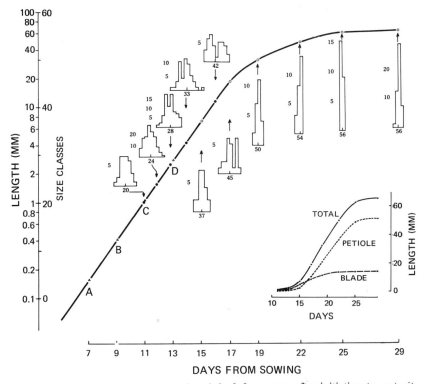

Fig. 5.2. Length–time relation for the fourth leaf, from soon after initiation to maturity. A–D mark the developmental stages of Fig. 5.1. The size distribution histograms given above (or below) the curve are for the times indicated by arrows. The histograms enclose a constant area, frequencies are to the left, and the size class which contains the median leaf is given below in each case. The inset shows the length–time relation for the blade, petiole, and total leaf on an absolute scale. (Williams and Rijven, 1970, Fig. 3.)

distribution of lengths on days 16 and 17 – a phenomenon which one might well dismiss as a chance one, were it not for a related condition in wheat (Fig. 5.7). After emergence there is a rapid reduction in relative length variation, with a range of only 54 to 74 mm on day 29. The lengths given in column 2 of Table 5.1 are derived from Fig. 5.2.

Because of the large range of primordium and leaf size in this study, two methods were used for cell counting. These involved direct counting of nuclei in transverse sections, and counting after tissue maceration (see the Appendix). Cell numbers are plotted on a logarithmic scale in Fig. 5.3, and there is a very satisfactory agreement between the two methods where the values overlap. From day 7 to day 13 the results are

165

Trifolium subterraneum

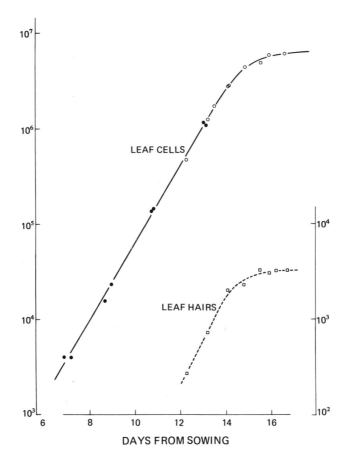

Fig. 5.3. Leaf cells and leaf hairs of the fourth leaf as a function of time. Values based on nucleolar counts (●) and on the maceration technique (○). The right-hand scale refers to leaf hairs only. (Williams and Rijven, 1970, Fig. 4.)

adequately described by a straight line, and the slope of this line implies a mean generation of time 18 h. Extrapolation of the curve at its upper end suggests a total of about 6.5 million cells for the fourth leaf.

The rather low figure of 0.197 pg DNA-phosphorus per cell was found for these clover leaves. This mean value is based on remarkably uniform estimates for several stages of growth and was used to convert the cell numbers of Fig. 5.3 to the DNA dry matter of Fig. 5.4. Wheat has 4.94 pg DNA-phosphorus per cell, which is 25 times that for the clover cell.

Trifolium subterraneum

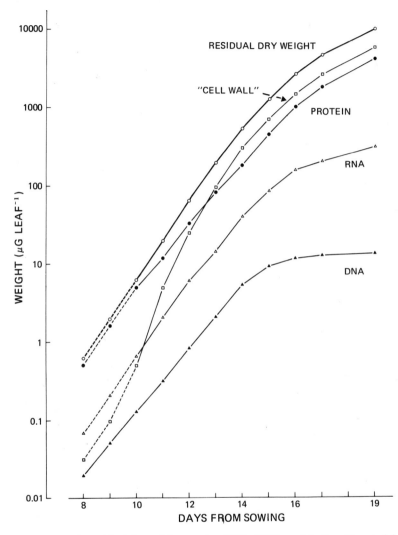

Fig. 5.4. Changes in residual dry weight proteins, RNA, DNA, and 'cell wall' materials of the fourth leaf, expressed as dry matter per leaf. (Williams and Rijven, 1970, Fig. 5.)

The growth of leaves from their initiation through to maturity, even with quite small leaves such as those of flax, covers such a tremendous size range that it is usually necessary to integrate either volume and dry weight (Figs. 4.2.9 and 4.7.7) or volume and fresh weight (Figs. 4.1.15, 4.5.8, 4.6.13, 4.6.14 and 4.8.11) to cover the whole range.

Table 5.1. *Growth of the fourth leaf of subterranean clover*
(quantity per leaf)

Days from sowing	Length (mm)	Volume (nl)	Fresh weight (mg)	Dry weight (mg)	Residual dry weight (mg)
7	0.16	1.09	—	—	—
8	0.25	3.20	—	—	—
9	0.40	9.39	—	—	—
10	0.66	27.6	—	—	0.0062
11	1.05	81.1	—	—	0.0193
12	1.72	238	—	—	0.0643
13	2.8	700	1.09	0.245	0.194
14	4.5	2058	—	—	0.531
15	7.2	6048	9.50	2.02	1.262
16	11.8	—	—	—	2.69
17	19.0	—	41.6	7.61	4.67
19	32.4	—	82.8	14.36	10.35
22	49.8	—	132.4	23.88	15.47
25	62.6	—	157.5	31.20	19.90
29	65.5	—	171.1	41.27	26.49
36	65.4	—	169.7	38.15	26.42

Weighing of any kind is ruled out for very early stages because one is there concerned with the rather arbitrary, but all-important definition of boundaries between primordium and stem. These are more objectively made by volume integration and a little mensuration than they are with a scalpel. If the reader is not convinced of this, let him try to visualize the difficulties of dissecting out that portion of a tunica layer which is destined to become a leaf primordium (cf. Fig. 4.1.3). Theoretically one could use volume throughout, were it not for the technical difficulties of determining volumes of older tissues with complex systems of intercellular spaces. The data of Table 5.1 are of special interest here because they include overlapping values for volume and weight. In fact, the common points used for this leaf in Fig. 4.5.8 are those for day 15, and these yield an apparent density of 1.571 (g fresh weight per ml). The parallel figure for day 13 is 1.557, which is reassuring for the procedure. However, a little calculation will show that a volume–dry weight link would have been almost as good. A further convention which is probably worth observing is to make the link at or very near leaf emergence. In our experience, intercellular spaces are negligible to that point, and most of the really important changes in form and rate – developmentally speaking – have already occurred.

The volumes of column 3 of Table 5.1 are derived from the regression

Trifolium subterraneum

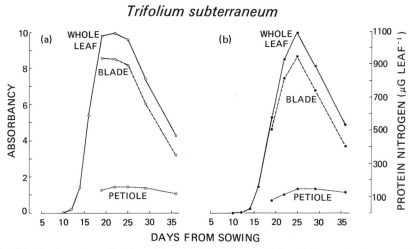

Fig. 5.5. Absolute change in (a) nucleic acids (absorbancy units) and (b) protein nitrogen of the whole leaf, blade and petiole of the fourth leaf. (Williams and Rijven, 1970, Fig. 7.)

for the fourth leaf, in Fig. 4.5.8. They exclude the leaf hairs, which become conspicuous after day 11. The residual dry weights for days 10–17 and much of the chemical data of Fig. 5.4 are based on the concept of age equivalence and on short-term linear regression (see the Appendix), but those for days 19–36 are direct estimates for samples harvested on the days indicated.

The differences between the dry weights and residual dry weights of Table 5.1 represent all those leaf constituents soluble in the exhaustive series of extractants listed by Williams and Rijven (1965, p. 723). There is little doubt that the main constituents are carbohydrates, lipids, minerals, and nitrogenous substances, though it was not anticipated that these could account for 21–39 per cent of the total dry matter. The residual dry weight is believed to be made up almost solely of proteins, the nucleic acids, and cell–wall materials, with some contamination of the latter with starch after leaf emergence. The time course for these constituents up to day 19 is shown in Fig. 5.4, and as a percentage distribution diagram right to maturity in Fig. 5.6. What happens to the proteins and nucleic acids after day 19 is best shown by the absolute data of Fig. 5.5. Clearly there is a massive breakdown of nucleic acids, especially in the leaf blade, and mainly of RNA, since DNA-phosphorus constituted only 5 per cent of the total nucleic acid phosphorus on day 19. The peak content of protein–nitrogen occurs on day 25 and perhaps a few days earlier in the blade than in the petiole.

169

Trifolium subterraneum

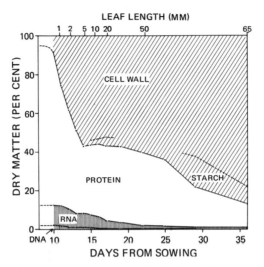

Fig. 5.6. Changes with time in the percentage distribution of dry matter between DNA, RNA, protein and 'cell wall' materials in the fourth leaf. Some starch was present in the cell wall fraction after day 15. Leaf lengths are indicated above, 65 mm being the final length. (Williams & Rijven, 1970, Fig. 6.)

Returning to Fig. 5.4 it will be noticed that this includes some rather fanciful extrapolations to day 8 (but see Williams and Rijven, 1970, p. 160 for their justification). These certainly help one to gain some insight into the events leading to the dramatic changes in composition from day 10 on (Fig. 5.6). The data show that cell wall materials increase from 8 to 57 per cent of the residual dry weight during a period of four days, and that this rise is almost matched by a fall from 79 to 35 per cent in the protein component. The reasons for these changes will be examined further below.

Changes in quantities per cell are recorded in Table 5.2, and an independent estimate for the dome of a nine-day apex suggests that mean cell volume has a starting point of 0.223 pl, which is equivalent to $(6.1 \, \mu m)^3$. Within the leaf primordium there is a steady increase to 0.823 pl on day 14 followed by a more rapid increase before steadying to a mature mean size of 20 pl per cell or $(27 \, \mu m)^3$. RNA-phosphorus rose fairly steadily to a maximum of 4.55 pg per cell on day 19; the subsequent fall assumes no loss of DNA-phosphorus. Protein-nitrogen per cell, on the other hand, is remarkably constant for days 10–14; it rises to a high value on day 25, and then falls sharply again.

Table 5.2. *Growth of the fourth leaf of subterranean clover*
(quantity per cell)

Days from sowing	Volume* (pl)	RNA-phosphorus (pg)	Protein-nitrogen (pg)	Cell wall (pg)
7	0.290	—	—	—
8	0.335	—	—	—
9	0.388	—	—	—
10	0.449	0.97	12.7	8.0
11	0.520	1.19	12.3	31.6
12	0.601	1.38	13.3	62.1
13	0.697	1.28	13.2	94.0
14	0.823	1.45	11.8	90.6
15	1.341	1.73	16.3	157.2
16	—	2.54	28.6	264
17	5.26	3.01	47.1	430
19	10.35	4.55	102.0	918
22	16.43	4.22	143.1	1437
25	19.17	4.06	167.8	1966
29	19.70	3.11	136.6	3185
36	19.98	1.72	82.2	3523

* Excluding the leaf hairs. Values for days 17–36 are based on leaf water.

The tremendous overall increase, more than 400-fold in the amount of cell wall material per cell, calls for a digression on observable events. Leaf hairs were first noted on day 9 as small papillose processes towards the distal ends of the folioles, and they were longer and much more numerous on day 11. Thereafter the actual numbers per leaf are recorded in Fig. 5.3. Under a dissecting microscope the primordia change in appearance from the naked condition of Fig. 4.5.2 to one in which the epidermis, just before leaf emergence, is quite hidden by some 3000 parallel hairs of mean length 1.1 mm. When preparing the drawings of Fig. 5.1 the hairs were necessarily ignored. That the mature leaf is rather sparsely coated with hairs is due to a 40-fold increase in leaf area after all the hairs have matured. Williams and Rijven (1970) calculate that a maturing leaf hair has a volume which is about 500 times that of a leaf cell on day 14, and a correspondingly large internal surface area upon which cell wall materials are being deposited.

Hair growth, however, is not the only phenomenon contributing to the rapid increase in cell wall materials during days 10–14. Williams and Bouma (1970) note the presence of a few mature phloem elements in the three main vascular bundles on day 10, and transverse sections show that these bundles have reached an advanced stage of differen-

Table 5.3. *Growth of the fourth leaf of wheat*
(quantity per leaf)

Days from sowing	Length (mm)	Volume (nl)	Residual dry weight (mg)	$10^{-3} \times$ No. of cells	DNA-phosphorus (μg)	RNA-phosphorus (μg)	Protein-nitrogen (μg)	Cell wall (μg)
7	0.24	5.6	—	—	—	—	—	—
8	0.31	8.9	—	—	—	—	—	—
9	0.39	14.2	—	—	—	—	—	—
10	0.51	22.2	0.0054	—	—	—	—	0.65
11	0.68	36.7	0.0082	13.8	0.068	0.105	0.901	0.98
12	1.01	70.1	0.0172	21.7	0.107	0.196	1.821	2.07
13	1.60	145	0.0381	46.6	0.230	0.420	4.23	4.76
14	3.16	394	0.0899	105.3	0.520	1.183	9.37	13.33
15	7.75	1292	0.229	260	1.282	2.76	21.0	55.1
16	19.5	3751	0.569	512	2.53	6.36	47.5	178
17	46.9	—	1.387	854	4.22	12.07	108.9	534
18	80.3	—	3.327	1413	6.98	21.8	241	1516
21	176	—	13.99	2935	14.5	60.6	938	7334
25	284	—	32.56	3704	18.3	86.6	2162	17940
31	276	—	38.4	3259	16.1	58.9	2465	22200
39	274	—	38.4	3381	16.7	43.0	2282	23510

tiation by day 14, including several metaxylem vessels and massive caps of non-lignified mechanical tissue (Fig. 4.5.3). It is not yet possible to assess the relative contributions of hair development and vascular differentiation to the rapid increases in cell wall materials.

Post-emergence increases in cell wall materials as such are somewhat obscured by the presence of starch, but it is clear that they increase about ten-fold over days 16–29. This, of course, is the main period of cell expansion and presumably of further cell wall thickening in all tissues.

5.2 WHEAT

The general characteristics of leaf growth in wheat have been discussed at some length already, and the main features are summarized in Fig. 4.7.14. This summary, and the drawings of Fig. 4.7.13 should be kept in mind as a background to the details of chemical change which follow. The fourth leaf was chosen because its primordium is the first to be initiated after germination, and because its growth is little influenced by seed reserves.

The length–time relation of Fig. 5.7 and Table 5.3 is more complex than that for subterranean clover (Fig. 5.2), and is certainly not expo-

Triticum aestivum

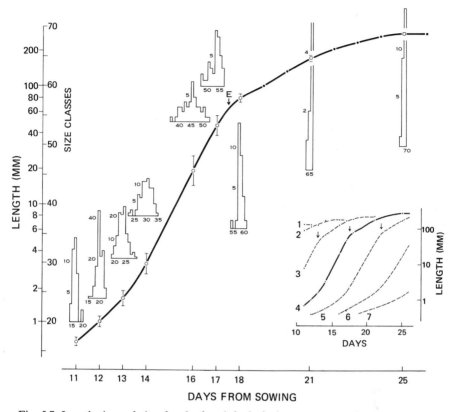

Fig. 5.7. Length–time relation for the fourth leaf of wheat. Means at times of sampling are shown with upper and lower quartile values. The other values are means for plants remaining after leaf emergence, E. The size distribution histograms given above (or below) the curve enclose a constant area, frequencies are to the left and the size classes are given below. The inset shows the length–time relations for leaves 1–7 for the same period. (Williams and Rijven, 1965, Fig. 3.)

nential throughout pre-emergence growth. Relative length growth is rather slow to day 13, and then increases to about double that rate for the few days before emergence. It decreases abruptly to a low rate following emergence and falls to zero by day 25. The inset to Fig. 5.7 shows that the pattern of length growth is essentially the same for successive leaves, though the rates for corresponding phases tend to decrease and their durations to increase with leaf number.

The size distribution histograms of Fig. 5.7 again show a high correlation between the rate of growth and the range of size encountered

173

Triticum aestivum

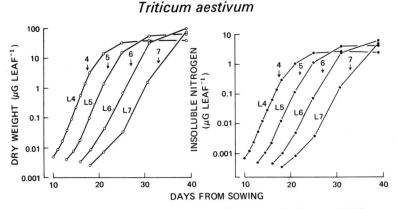

Fig. 5.8. Changes in residual dry weight and insoluble nitrogen for leaves 4–7. The arrows indicate times of leaf emergence. (Williams and Rijven, 1965, Fig. 5.)

at any time. Length variation is not excessive at day 11, but increases to a maximum on day 16 (with a range of 7–44 mm). The reduction in variation after leaf emergence is quite dramatic and yields a range of only 250–315 mm at maturity. It was this pattern of variation which prompted the setting up of age-equivalence procedures, especially for the study of chemical change in rapidly growing organs.

For the purposes of this summary it seemed better to present the absolute data for wheat in tabular form (Table 5.3) rather than as text figures, especially as there are some significant alterations to the data as originally presented by Williams and Rijven (1965). In a note added in proof, these authors supply a much improved estimate of the DNA-phosphorus content per cell – 4.94 ± 0.176 pg. This value was used to calculate the cell numbers of Table 5.3, and hence the quantities per cell in Table 5.4. It will be noted that a leaf which is only 0.68 mm in length already has a complement of 14 000 cells, and that the mature leaf has 3.4 million cells.

Only residual dry weights are given in Table 5.3, but other work indicates that about 20 per cent of the original dry matter had been removed by the extractants. There was no evidence of contamination with starch, and the residual dry weight is believed to be made up almost solely of proteins, the nucleic acids, and cell wall materials.

The original purpose of the study was to define in chemical terms the established changes in pre-emergence leaf growth, so it was disappointing to find that the new information did not extend very far back into the early phase of slow exponential growth. To remedy this short-

174

Triticum aestivum

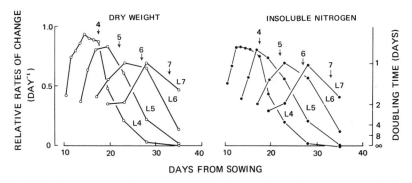

Fig. 5.9. Relative rates of change in dry weight, R_W, and insoluble nitrogen, R_{IN} for leaves 4–7. The arrows indicate times of emergence and the scales of R and of doubling time apply to both sets of data. (Williams and Rijven, 1965, Fig. 8.)

coming, estimates of residual dry weight and of insoluble nitrogen were obtained for leaves 5, 6 and 7 of the same experiment (Fig. 5.8). The results show without a doubt that both quantities increase more slowly at first. This is more effectively shown in Fig. 5.9 in terms of relative rates of change in dry weight, R_W, and in insoluble nitrogen, R_{IN}, for leaves 4–7. For both these rates there is good evidence for relatively low initial values for each leaf, though only with leaf 7 has the low value been shown to continue for any length of time.

Table 5.3 shows that length growth and DNA synthesis cease by day 25, that dry weight and the proteins increase to day 31, and that there is a considerable loss of RNA after day 25.

The constituents of the residual dry weights are expressed as percentages of their combined weights in Fig. 5.10 and, as for the clover leaf, there are spectacular changes in the few days prior to emergence. Between days 14 and 18, cell wall materials increase from 15 to 46 per cent, proteins fall from 65 to 45 per cent, and the nucleic acids together fall from 20 to 9 per cent. During maturation, the same trends continue so that the values become 61, 37, and rather less than 2 per cent at day 39 for cell wall, protein and the nucleic acids respectively.

Changes in the quantities per cell for the same constituents and for volume are set out in Table 5.4, and in considerable detail for RNA-phosphorus in Fig. 5.11. The volume per cell increases rather slowly until day 14. It then increases rapidly, presumably to very high values at maturity. Wright (1961), for instance, gives mean values of 500 pl for mature wheat coleoptile cells. Protein–nitrogen per cell is quite

175

Triticum aestivum

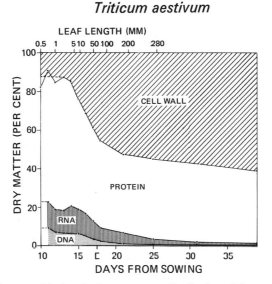

Fig. 5.10. Changes with time in the percentage distribution of dry matter between DNA, RNA, protein and cell wall materials in the fourth leaf. Leaf lengths indicated above, 280 mm being the final length. E, time of leaf emergence. (Williams and Rijven, 1965, Fig. 6.)

remarkably constant until day 16, but increases eight-fold by day 31. However, histological examination of the material suggests that the slow early changes in volume and protein–nitrogen may represent a balance between continued enlargement of the cells of the epidermis and mesophyll, and the progressive differentiation of many small cells in the provascular strands. By contrast, Table 5.4 shows that RNA-phosphorus and cell wall material per cell increase from the first observation. Since the wheat leaf has no hairs, the pre-emergence rise in cell wall materials can only be attributed to vascular differentiation and wall thickening. Later, the increasing cell size, and the progressive appearance of metaxylem and sclerenchyma would all contribute. Fig. 5.11 shows that the time course for RNA-phosphorus per cell differs markedly for different parts of the leaf. Thus the distal half, and the second quarter of the blade reach their maxima at or before day 21. The proximal quarter of the blade has a high maximum at day 25, and the leaf sheath value is still rising between days 25 and 30. Clearly leaf growth and development in wheat, and perhaps in all grasses, progresses fairly slowly from the tip, through the blade and sheath, and eventually into the internode below, as was suggested long ago by Sharman (1942).

176

Triticum aestivum

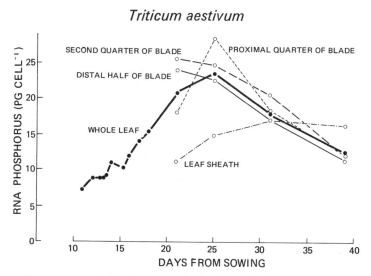

Fig. 5.11. Changes in RNA phosphorus per cell in the fourth leaf. From day 21, separate values are shown for four parts of the leaf. (Williams and Rijven, 1965, Fig. 7.)

Table 5.4. *Growth of the fourth leaf of wheat*
(quantity per cell)

Days from sowing	Volume (pl)	RNA-phosphorus (pg)	Protein-nitrogen (pg)	Cell wall (pg)
11	2.67	7.63	65.4	71
12	3.24	9.05	84.1	96
13	3.11	9.02	90.9	102
14	3.74	11.24	89.0	127
15	4.98	10.64	80.9	212
16	7.32	12.38	92.7	347
17	—	14.13	127.5	625
18	—	15.43	171	1073
21	—	20.65	320	2499
25	—	23.38	584	4843
31	—	18.07	756	6812
39	—	12.72	675	6954

5.3 RELATIVE RATES OF CHANGE, *R* AND *G*

One of the really rewarding things about quantifying a range of attributes within a growing system is that many interrelationships – sometimes quite unexpected ones – can be explored with the aid of quite simple arithmetic and an electronic desk calculator. I certainly did not

Triticum aestivum

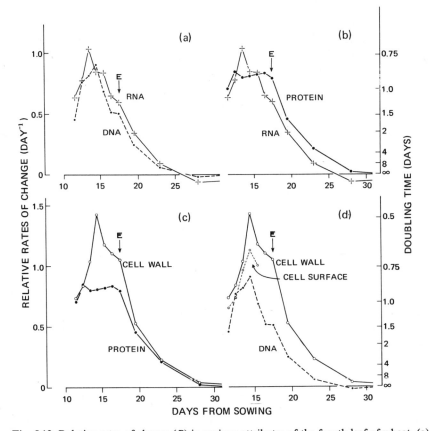

Fig. 5.12. Relative rates of change (R) in various attributes of the fourth leaf of wheat. (a) Values for DNA and RNA (R_{DNA}, R_{RNA}). (b) Values for RNA and protein nitrogen (R_{RNA}, R_{PN}). (c) Values for protein nitrogen and cell wall materials (R_{PN}, R_{CW}). (d) Values for DNA, cell wall materials, and cell surface (R_{DNA}, R_{CW} and R_{CS}). E, time of leaf emergence. (Williams and Rijven, 1965, Fig. 10.)

anticipate for instance, that the wheat data could be made to yield estimates of the rate of deposition of cell wall material per unit of cell surface. But such proved to be the case. As for the calculator, even our old friend R can be quite daunting to some, until it is discovered that values can be *done* in less than 10 seconds on a machine with a ln stop.

The relative growth rate concept can be particularly valuable here because rates and doubling times can be compared at a glance, and in many combinations. This has been done in Figs. 5.12 and 5.13 for wheat and clover respectively. Such sets of graphs summarize the dynamics

178

Trifolium subterraneum

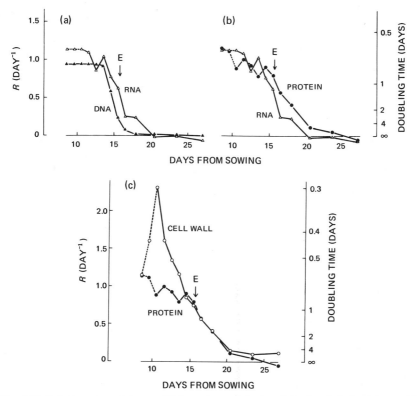

Fig. 5.13. Relative rates of change (R) in various attributes of the fourth leaf of clover. (a) Values for DNA and RNA (R_{DNA}, R_{RNA}). (b) Values for RNA and protein nitrogen (R_{RNA}, R_{PN}). (c) Values for protein nitrogen and cell wall material (R_{PN}, R_{CW}). E, time of leaf emergence. (Williams and Rijven, 1970, Fig. 8.)

of the systems in a very effective way. Thus, since R_{DNA} values can be read as mean cell generation times, we see at once that a minimum value of approximately 18 h occurs on day 14–15 for wheat, and that this is also the value for pre-emergence growth in clover.

Other derivatives obtainable from the data are rates of production of one component of the system per unit of another component. Thus it is meaningful to determine the rate of production of RNA per unit of DNA, always remembering that the result will be a mean value for what is almost certainly a very variable quantity as between tissues. The instantaneous rate of production, G, of RNA per unit of DNA is given by

$$G_{(RNA, DNA)} = \frac{1}{DNA} \cdot \frac{d(RNA)}{dt},$$

179

Table 5.5. *Rates of production,* G *of one component of the leaf per unit of another component in subterranean clover*

Interval (days)	$G_{(RNA, DNA)}$ (day^{-1})	$G_{(PN, RNA)}$ (day^{-1})	$G_{(CW, PN)}$ (day^{-1})	$G_{(CW, CS)}$ (ng mm^{-2} day^{-1})
10–11	6.5	5.8	0.56	117
11–12	7.4	5.7	0.95	182
12–13	6.0	5.3	1.29	224
13–14	7.5	4.1	1.64	234
14–15	6.4	4.6	1.36	177
15–16	6.9	4.7	1.13	157
16–17	3.6	4.4	0.83	127
17–19	4.6	4.4	0.58	111
19–22	—	1.8	0.23	52
22–25	—	1.1	0.18	43
25–29	—	—	0.32	70
29–36	—	—	0.07	—

PN, protein nitrogen; *CW*, cell wall material; *CS*, cell surface, including leaf hairs.

and the mean value for a finite time interval may be calculated from the formula:

$$G_{(RNA, DNA)} = \frac{\log_e DNA_2 - \log_e DNA_1}{t_2 - t_1} \cdot \frac{RNA_2 - RNA_1}{DNA_2 - DNA_1}.$$

The error involved in the use of this formula is negligible for short time intervals, and when the relation between the variables is approximately linear.

The mean value of $G_{(RNA, DNA)}$ for wheat, calculated from the dry weight equivalents, is 1.97 day^{-1} between days 13 and 18, but it is only 0.80 day^{-1} during early exponential growth. Similarly the rate of production of proteins per unit of RNA, $G_{(PN, RNA)}$ is 3.92 day^{-1} for the interval days 11–21 but only 2.45 day^{-1} during early exponential growth. Later in their paper Williams and Rijven (1965) indulge in a rather tortuous analysis which leads to various estimates of the rate of deposition of cell wall material per unit of internal cell surface – a figure of 50 ng mm^{-2} for $G_{(CW, CS)}$ is suggested as meaningful. The analysis is of small consequence in itself because of the rather wild assumptions which are made, but it surely points to the potential of the concept for the in-vivo study of chemical change in growing systems.

Table 5.5 is a full statement of trends in four kinds of G for the clover leaf study. In three of them the components are expressed in common units (dry weight), so the rates become relative increments per day.

$G_{(RNA, DNA)}$ is constant over days 10–16, with a mean value of 6.8

day^{-1}. Then follows a brief period of activity at a lower level, but there is no net synthesis of RNA after day 19.

The values of $G_{(PN,\,RNA)}$ fall into three groups, with mean values of 5.6, 4.4 and 1.5 day^{-1} for days 10–13, 13–19 and 19–25 respectively. It is tempting to see in the three levels a hint that the synthesis of specific proteins or groups of proteins may be dominant for successive stages of development.

By contrast, the values of $G_{(CW,\,PN)}$ rise to a maximum of 1.64 day^{-1} for the interval days 13–14. The subsequent steady fall in the rate is interrupted only by a high value for days 25–29, and this could be due to starch accumulation. The parallel set of values for $G_{(CW,\,CS)}$ are believed to be soundly based, and allow for the contribution of the leaf hairs to total internal cell surface. That these mean rates should show high values throughout the pre-emergence phase of leaf hair growth and vascular differentiation is consistent with the earlier observation of very rapid increases in cell wall material per cell at this stage. That so few cells could so significantly increase the mean rate of cell wall deposition can only mean that the cells are very large (the hairs) or have very thick walls (hairs and vascular elements). Histological examination amply supports this conclusion.

The foregoing quantitative descriptions of chemical change in typical leaves of subterranean clover and wheat have shown many similarities. Proteins account for over 80 per cent of the young primordium in clover, and a little less than 70 per cent in wheat. Cell wall materials and DNA are less prominent in clover than in wheat, but RNA is present in similar amounts. Both plants exhibit dramatic changes during the few days prior to leaf emergence, with cell wall materials rising to 50 per cent and more of the total. This is when the bulk of cellular differentiation takes place. After emergence there is a steady increase in the proportion of cell wall materials at the expense of proteins and nucleic acids in both wheat and clover.

One curious feature of the data of both leaf types is that, during the period of rapid differentiation, protein–nitrogen per cell is constant in spite of a steady increase in mean cell size. With this increase there is rapid vacuolation and correlated changes in surface–volume ratios, especially in vascular elements and leaf hairs. One might well ask how far such changes provide a basis for the high rates of cell wall deposition taking place in these cells. Incidentally, the maximum value of $G_{(CW,\,PN)}$ for clover is 1.64 day^{-1} compared with 0.94 day^{-1} for

181

wheat. This difference can reasonably be attributed to the hairs of the clover leaf.

The five-day period before leaf emergence is also that of most rapid growth in both wheat and clover. Mean values of R_{RNA} are 0.82 day^{-1} and 0.87 day^{-1} respectively, and those for R_{PN} are 0.82 day^{-1} and 0.89 day^{-1}. However, this rather remarkable agreement between the species does not extend to the G values. Those for $G_{(PN, RNA)}$, for instance, are 6.90 day^{-1} and 4.86 day^{-1} for wheat and clover respectively. The contrast is even more startling with $G_{(RNA, DNA)}$ which has mean values of 1.82 and 6.84 day^{-1} for wheat and clover respectively. In this case it may be argued that the wheat DNA is much less effective in producing the required RNA for a comparable level of general metabolism, and that the species difference could be ascribed to the fact that wheat is hexaploid and subterranean clover is diploid. Clearly, however, such interpretations are so wildly speculative that there is little merit in indulging in them. They are offered here primarily to illustrate the potential of the approach.

The detailed description of a plant or a representative leaf under one set of conditions is not an obviously rewarding task. However, if a wide range of attributes are included in the study, meaningful patterns and interrelations emerge. The task ahead might well include the use of these techniques for the analysis of treatment effects within shoot-apical systems. Such information will be essential to a deeper understanding of the nature of biological order in plants.

6 The growth of an inflorescence

This short chapter is concerned solely with the growth and development of the ear of wheat, *Triticum aestivum*. At the outset it is freely admitted that the work was undertaken largely in response to the challenge of describing such a complex biological system with precision. There was also the hope that such work would throw light on the nature of the transition from vegetative to reproductive development in the wheat plant. The chapter draws heavily upon the descriptive paper by Williams (1966*a*) and to a lesser extent on the more general statement on the inflorescence in Gramineae (Williams, 1966*b*). Studies of the developmental morphology of the wheat spike have been made by Bonnett (1936) and Barnard (1955), and the latter's histogenetic classification of its members into foliar and cauline types is adopted.

The unit of the gramineous inflorescence is the spikelet, a group of one or more flowers – usually called florets – with a number of associated bracts. These are collected into inflorescences with a tremendous range of form, from the open panicle of *Avena* to very compact structures such as the spike in *Triticum*. A common form of the spikelet consists of an axis (the rachilla) bearing two sterile glumes at the base and an indefinite number of lemmas, each with a floret in its axil. In its mature state (at anthesis) the wheat spikelet is as depicted in Fig. 6.4E but its basic structure is more easily grasped at the developmental stage shown in Figs. 6.1 and 6.6. Fig. 6.1 shows two lateral spikelets in which the seventh (possibly the last) floret has just been initiated, and in which the glumes and other bracts have not yet grown up to conceal the inner members. The florets themselves display a beautiful array of developmental stages, and these can also be discerned in the transverse sections of Fig. 6.6. Those familiar with Arber's classic, *The Gramineae* (1934) will remember her profuse use of the transverse section as a means of demonstrating the range of floral structure to be found in the gramineous spikelet. The method is a powerful one, especially when used in conjunction with three-dimensional drawings and photographs.

Each floret, as we have seen, is subtended by a bract called a lemma (Fig. 6.7), and its first member is the palea, which is a two-keeled foliar structure closely resembling the prophyll of the vegetative tiller. The

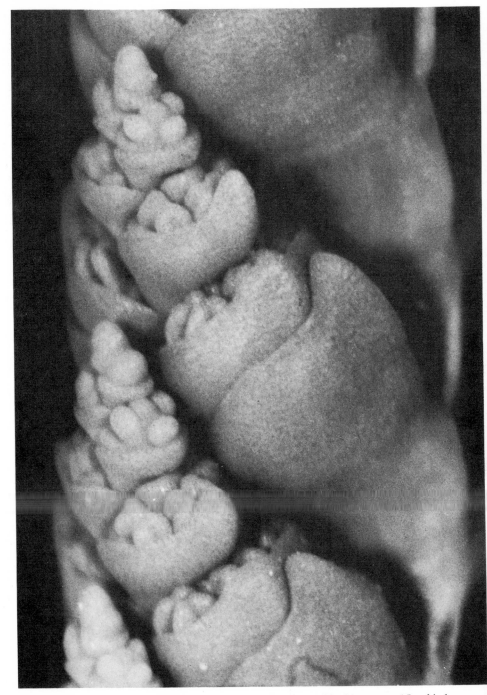

Fig. 6.1. Young inflorescence of wheat, 14 days after the double ridge stage of floral induction and 19 days before anthesis (36-day plant). (Williams, 1966*a*, Plate 1.) (×75.)

perianth proper is reduced to the lodicules, of which two are placed anteriorly. A third, posterior lodicule occurs in bamboos, and also in *Stipa*. There are three large stamens, which are exserted at anthesis by the very rapid growth of the slender filaments. The gynaecium consists of a unilocular ovary, usually interpreted as a single carpel – but see below. It is surmounted by two styles in wheat. Section C of Fig. 6.7 passes right through the ovary, and, because of its orientation within the carpel, this is seen in longitudinal section, and the large cell at its centre is probably the functional megaspore.

. The beautiful manner in which the floral parts fit together in the space available brings to mind Arber's speculation that pressure may play a part in moulding the form of floral parts. She suggests, for example, that the two keels of the palea are formed in the only spaces available, those between the lemma and the margins of the upper floret. She also quotes the classic speculation of Turpin (1819) that the dorsal lodicule is feeble in bamboos and absent in most grasses because of dorsiventral pressures between the developing floret parts. This is the first case known to the author of a proposal that physical constraint could be thought of as a causal mechanism for growth and development.

The developmental morphology of the wheat spike is set out pictorially in Figs. 6.2–6.4. This provides quite an accurate record because the axes photographed are medians from random arrays of 24, at two-day intervals, for days 20–38, and at four-day intervals thereafter. The plants were grown in a controlled environment with a temperature regime of $20°/15 °C$, and long days, using photoperiodic light during the whole 'night'.

In the 16-day vegetative apex of Fig. 6.2, leaf 6 is 0.4 mm long, leaf 7 is visible, and the apex is beginning to elongate prior to ear formation. Leaf initials are then formed quite rapidly but their further development is slight, and, under long-day conditions, the spikelet initials grow precociously. Only at the base of the spike do the foliar members remain recognizable for any length of time. Higher up the spike they are represented only by a crescent of periclinal cell divisions, with little or no effect on the form of the apex. There is, however, a curious reversal of this process once the terminal spikelet has been determined. This structure is entirely the product of the terminal apex and its prominent foliar bracts are systichous with the suppressed foliar members of which we have been speaking. That this is so is evident in Fig. 6.3. The 28-day apex of Fig. 6.2 has just initiated the

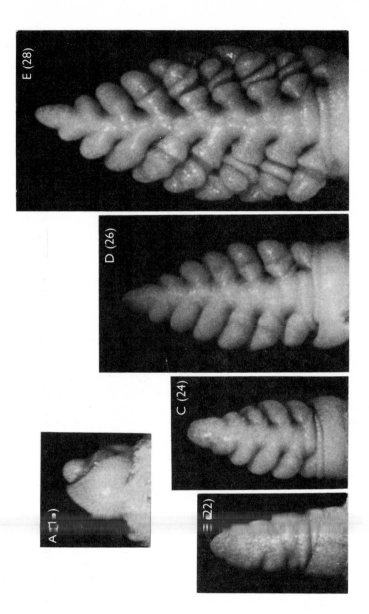

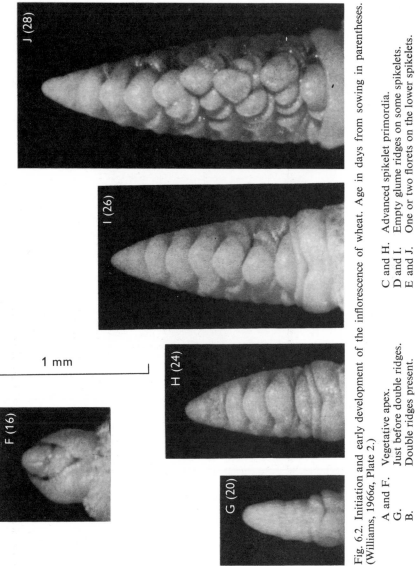

Fig. 6.2. Initiation and early development of the inflorescence of wheat. Age in days from sowing in parentheses. (Williams, 1966a, Plate 2.)

A and F. Vegetative apex.
G.   Just before double ridges.
B.   Double ridges present.

C and H. Advanced spikelet primordia.
D and I. Empty glume ridges on some spikelets.
E and J. One or two florets on the lower spikelets.

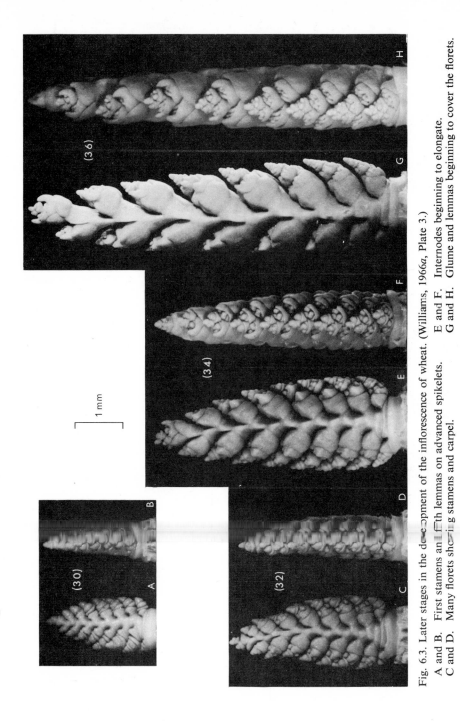

Fig. 6.3. Later stages in the development of the inflorescence of wheat. (Williams, 1966*a*, Plate 3.)

A and B. First stamens and fifth lemmas on advanced spikelets. E and F. Internodes beginning to elongate.
C and D. Many florets showing stamens and carpel. G and H. Glume and lemmas beginning to cover the florets.

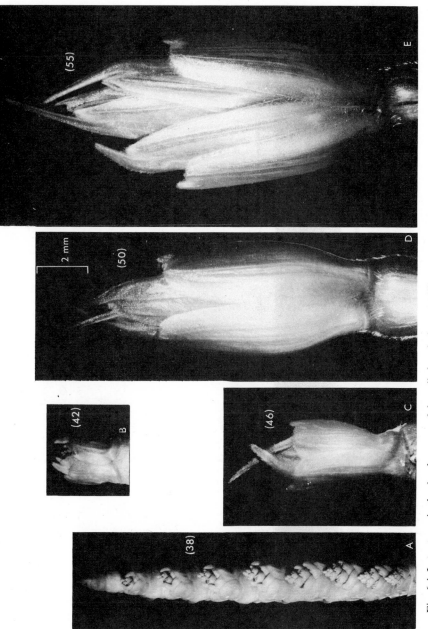

Fig. 6.4. Late stages in the development of the spikelet of wheat. (Williams, 1966a, Plate 4.)

A. 38-day inflorescence.
B. Florets almost completely covered by glumes and lemmas.
C. Awns present.

D. Glumes and lemmas fully grown.
E. On day of anthesis, glumes and lemmas spreading.

189

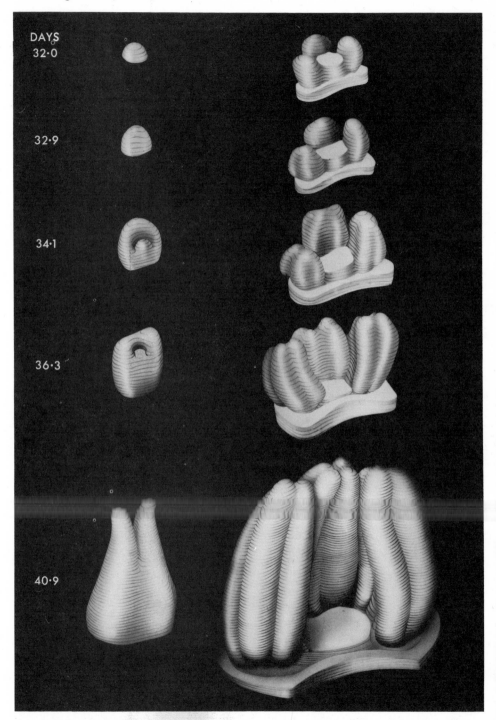

Fig. 6.5. Three-dimensional reconstructions of the carpel and stamens for five stages of development. Times in days from sowing are age equivalents based on length. (Williams, 1966a, Plate 5.) (×90.)

Triticum aestivum

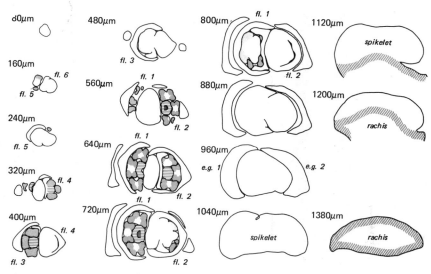

Fig. 6.6. Outline drawings of transverse sections spaced at 80 μm intervals from the tip of a median spikelet of a 36-day axis. Sections at 160, 320, 400, 560 and 720 μm pass through the carpels of florets 5, 4, 3, 2 and 1 respectively. The empty glumes are shown free in the section which is 960 μm from the tip, and the last three sections illustrate the union of the spikelet with the rachis. Anther loculi are stippled, and carpellary tissue is hatched horizontally. Ovules are present in florets 1 and 2. (Williams, 1966a, Fig. 1.) (× 25.)

Triticum aestivum

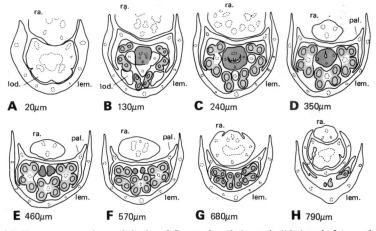

Fig. 6.7. Transverse sections of the basal floret of a 42-day axis (13 days before anthesis). Heights above junction of floret with rachilla are given in μm. Anthers (but not filaments) shown by stippling; carpel, style, and ovule are vertically hatched. lem. = lemma, lod. = lodicule, pal. = palea, ra. = rachilla. C, includes median section through ovule. D, below junction of styles, showing stylar canal. H, includes lemma and base of floret 3 between rachilla and upper parts of floret 1. (Williams, 1966b, Fig. 1.) (× 21.)

sixteenth lateral spikelet, and the next visible member on the right should be a glume – not a spikelet.

There is clearly a need at this point to clarify the distinction between cauline and foliar structures as these terms are used by Barnard (1955). Concerning vegetative growth, Gregory (1956) pointed out that on the same apex two types of cell group arise, the one leading to the production of the leaf which is of limited growth, the other to bud formation and thus to an organ of unlimited growth, replicating the whole apex. Much earlier, Sharman (1945) made a careful histological analysis of the difference between leaf and bud initiation in the Gramineae, and showed that the leaf is of relatively superficial origin, whereas the stem incorporates deeper tissues. On this basis Barnard demonstrated conclusively that the spikelet is cauline in origin, though obviously not an organ of unlimited growth. He classifies the following structures as foliar: foliage leaves, foliar ridges subtending the spikelets, glumes, lemmas, paleas, lodicules, the carpel and the ovary integuments. The following are cauline in origin: tiller buds, spikelets, florets, stamens and the ovule.

Readers are referred to Barnard (1955) for the histogenetic evidence, but the morphology of spikelet development is clear in Fig. 6.2–6.4. The glumes and the first and second lemmas differentiate successively as lateral ridges which more than half encircle the spikelet axes. As a third lemma is forming, a floret primordium becomes visible in the axil of the first lemma. A narrow ridge along the posterior side of the floret primordium initiates the palea. Then, on the oval cushion of the floret, two lateral stamens appear, followed quickly by the anterior stamen and the carpel. The lodicules are the last of the floral organs to become clearly visible. Later stages in the morphogenesis of the stamen and carpel are shown in the three-dimensional reconstructions of Fig. 6.5. These are for basal florets of median lateral spikelets. The stamens show a steady increase in size and in complexity of form. The carpel, however, passes through a complex sequence of changes in form, including the apparent obliteration of the apical dome by the anterior member of the carpel. This phenomenon would seem to have a parallel in the difficulty of distinguishing between the apex and the leaf primordia in the young tobacco seedling (Fig. 4.2.2). The apparent reconstitution of the apex on day 34.1 (Fig. 6.5) also has at least a superficial resemblance to an early event in tiller bud initiation in wheat (Fig. 4.7.3, day 10.3), where a large and very much flattened apex seems quite suddenly to differentiate into a prophyll initial and a residual

192

Triticum aestivum

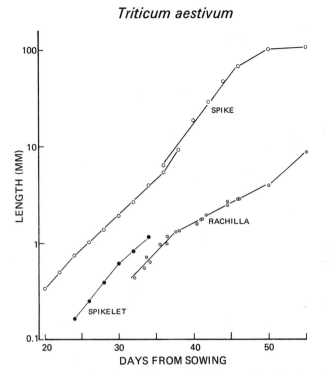

Fig. 6.8. Length growth of the spike, spikelet, and rachilla on a logarithmic scale. (Williams, 1966*a*, Fig. 8.)

apex. It is just possible, I suppose, that the fact that both apices are firmly in contact with more mature surfaces – the lemma above (Fig. 6.1), or the sheath of the subtending leaf (Fig. 4.7.3 and text comment) – could explain the similar appearances. Be that as it may, differential growth in the vicinity of the carpel apex causes this to be invaginated within the body of the carpel, where it differentiates as the ovule (Fig. 6.7C). Quite late in the development of the carpel, the twin styles appear as lateral processes and grow rapidly (Fig. 6.5). From a later analysis, Barnard (1957) concluded that the gynaecium is made up of as many as four foliar structures fused together – an anterior unit, a posterior unit and then two lateral units, the styles.

With this rather elaborate account of the developmental morphology of the wheat ear as background, we now turn to the length, volume and weight data recorded by Williams (1966*a*).

Length measurements of the spike and floral parts are presented in Figs. 6.8–6.10. The discontinuities of Fig. 6.8 relate to the conventions

Triticum aestivum

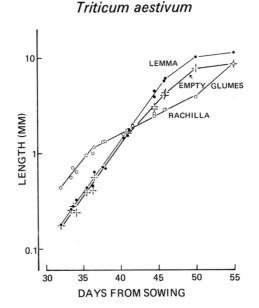

Fig. 6.9. Length growth of the rachilla, empty glumes, and lemma of a basal floret of a median spikelet. (Williams, 1966a, Fig. 9.)

of measurement, for, though spikelet and rachilla length are essentially of the same object, the first is taken to the centre of the rachis and the second to its junction with the rachis. In the commentary which follows, growth rate means relative length growth, R_L, or the slopes of the lines in the text figures.

The growth rate of the spike (Fig. 6.8) decreases slightly on day 24, and this is correlated with the onset of active spikelet growth (Fig. 6.2C). Spikelet growth is checked in turn on day 30, perhaps with the onset of floret growth (Fig. 6.3A). From day 36 onward, however, internode growth results in a sharp increase in the growth rate of the spike (Fig. 6.3G and H) and this is correlated with a further reduction in the rate of spikelet growth (= rachilla). The growth rate of the spike is then fairly constant until day 46 and ceases on day 50. There is then a sharp increase in the rate of rachilla growth, due also to internode extension.

The data for the rachilla are repeated in Fig. 6.9 together with those for the empty glumes and the lemma. These foliar organs all grow at a high rate for some days, but they do not show an early phase of slow growth such as occurs in the foliage leaf (Fig. 4.7.14). Indeed they do not

194

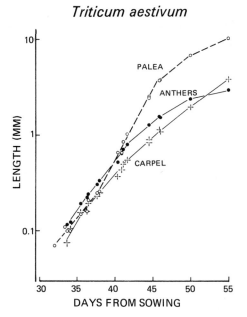

Fig. 6.10. Length growth of the palea, anthers, and carpel (including style) of a basal floret. (Williams, 1966*a*, Fig. 10.)

appear to be under constraint during early growth as are the leaves (see Figs. 6.1–6.4), and the flag leaf only loosely encloses the ear during this period. After day 46 the growth rates of the glumes and lemma fall away as did that of the spike. The rachilla is at first longer than the lemma of the basal floret, but is very much shorter by day 50. This is why the florets are at first naked on the rachilla, but become progressively covered by the lemmas and glumes.

Figure 6.10 records the length growth of the palea, anthers and carpel, and the palea *does* exhibit an early period of relatively slow growth (R_L values of 0.22 and 0.33 day^{-1} for days 32–38 and 38–46 respectively). The early period happens also to be one in which the palea is in close contact with neighbouring members of the system (Fig. 6.6), and this close contact no longer applies on day 42 (Fig. 6.7). Eventually the palea is almost as long as its associated lemma. Anther and carpel growth rates fall away fairly continuously with time.

Volume changes in the parts of the basal floret of a median spikelet are presented for the interval, days 32–46, and are linked with dry weight values for days 50 and 55 in Fig. 6.11. The link is a tenuous one for it rests on a single common value for the carpel on day 50. The

195

Triticum aestivum

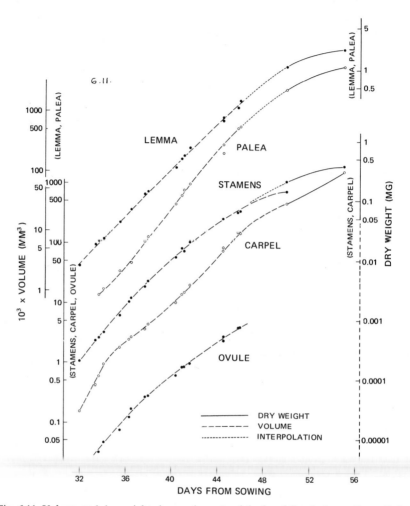

Fig. 6.11. Volume and dry weight changes in parts of the basal floret of a median spikelet. (Williams, 1966a, Fig. 12.)

volume for stamens for the same day was rejected as a link because of excessive shrinkage during processing. This text figure is based on a rather special use of the concept of age equivalence and will be discussed further in the appendix. The outcome gave very smooth sequences of values, and these were fitted with freehand curves.

The curves for the lemma, the palea, and the stamens are as might be expected from the length data reported above, but that for the carpel

Triticum aestivum

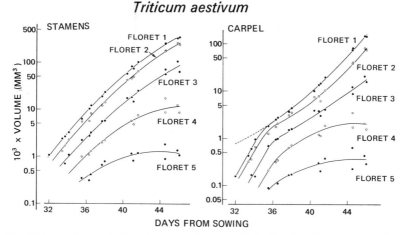

Fig. 6.12. Volume changes in the stamens and carpels of the first five florets of a median spikelet over the interval days 32–46. (Williams, 1966a, Fig. 14.)

is in striking contrast with expectation. It implies quite rapid changes in relative growth rate with time as follows:

Interval (days)	32–4	36–8	40–2	44–6	48–50
R_V	0.82	0.33	0.44	0.48	0.19

These changes find their readiest explanation in the form changes of Fig. 6.5, though there is little to be gained by pressing the point. It is clear, however, that the growth curve is dominated, first by the growth of the anterior carpel (days 32–34) and then by the growth of the lateral members or styles (with a maximum for days 44–46).

The contrast between the growth curves for stamens and carpels is also very evident in Fig. 6.12 which gives them for the first five florets of a median spikelet. A feature of both sets of data is the rapid departure from trend with increasing floret number. It would seem unlikely that florets 4 and 5 could produce mature anthers or carpels, and we know that median spikelets of this cultivar set only two or three grains of wheat.

At the beginning of the chapter the hope was expressed that the precise description of growth and development in the inflorescence might shed some light on the transition from the vegetative to the reproductive condition. The most obvious point that emerges is that there is a rather delicate balance between the foliar and cauline types of activity. In early vegetative growth the apex is completely dominated by foliar initials, but these are produced further and further from the

197

apical dome with increasing leaf number. When this process has proceeded far enough, and only then, the apex can elongate as a prelude to floral initiation. The main difference between long- and short-day plants in this respect is that the process is slower in short days, and many more foliage leaves are initiated on the primary shoot. The only additional cauline activity in the vegetative plant is the production of tiller bud primordia, and these quickly become foliar dominant.

As has been shown, new foliar members on an elongating apex are quickly suppressed by the precocious growth of axillary structures – the spikelet primordia. This yields the well known double-ridge condition and the sequence of events recorded in Fig. 6.2. Spikelet primordia clearly differ from tiller bud primordia in that their foliar members arise well back from the bare apex from the start, and never dominate it. It is a curious fact, nevertheless, that with each new order of branching in the inflorescence, foliar activity does reassert itself. Foliar activity is considerable in the glumes and lemmas of the first-order branch; it achieves even more normal expression in the palea of the second-order branch; and again in the lateral members of the carpel, if they are accepted as parts of a third-order branch. This last suggestion was made by Williams (1966b) on rather slender evidence.

While the foregoing discussion of the balance between foliar and cauline activity does not take us very far, it will be clear that foliar dominance within the apex is essential to the maintenance of vegetative growth. Only when foliar dominance is overcome can the full potentialities of the apex hope to gain expression, for the events which are most characteristic of reproductive development are all cauline in nature. Thus the beginnings of a biological understanding of floral initiation may lie in this direction, and one suspects that here, too, comparative studies will pay dividends.

7 The growth of wheat tillers

Long acquaintance with the wheat plant suggested that the genesis and growth of tiller buds might provide a suitable test system for the supposed effects of constraint on growth rates and on the genesis of form. Detailed accounts of collaborative work on this problem are currently being published (Williams *et al.* 1975; Williams and Langer, 1975), and a preliminary experiment involving both the removal of constraint and the application of additional constraint will be published by Williams and Metcalf (1975).

Drawings illustrating the initiation and early growth of the bud in the axil of leaf 2 are given in Fig. 4.7.3, and later stages for similar tillers appear as Figs. 7.1 and 7.2. The external shape of these early-stage tillers is entirely determined by their location. In Fig. 7.1 the tiller is shown in place and also removed at the level of its half-junction with the axis, thus revealing the snug cavity in which it is growing. In Fig. 7.2 the moulding is even more obvious, with a vertical dent to the right marking the position of the midrib of leaf 3. Note also that the first tiller leaf has the appearance of being extruded from between the margins of the prophyll, and a small piece of the inner margin of the leaf is actually caught between the outer margin and the prophyll.

The three-dimensional drawings are a little misleading visually because the removal of the relatively massive and slightly more mature tissues of the outer leaf sheath gives a false impression of freedom. The transverse sections of Fig. 7.3 (also Fig. 4.7.6 above) will counter such an impression; they also illustrate some additional features of the system. It will be noted for instance that tiller buds are located to one side or the other of the mid-bundle of the axillant leaf, and they are usually all on the same side of a line joining the mid-bundles of successive leaves. For instance, T1, T2 and T3 are to the left of such a line in Fig. 7.3. This feature tends also to be linked with the under- and over-relationship of the margins of the inner leaf. We will return to these phenomena in a later section on the effects of stress on the genesis of form.

The three sections of Fig. 7.3 also illustrate three stages in the genesis of tiller 3. These stages are roughly equivalent in appearance to those which can be visualized by interpolation at days 4, 6 and 8 in the

Triticum aestivum

TILLER 4

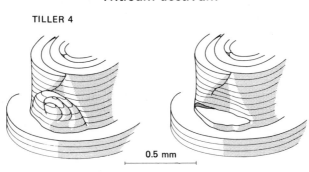

DAY 20

Fig. 7.1. Three-dimensional reconstruction of the fourth tiller bud, and of the cavity into which it fits. Leaf 3 has been removed just above its junction with the stem. The contours are 40 μm apart.

Triticum aestivum

TILLER 3

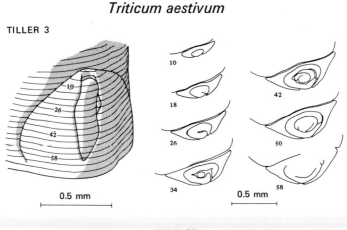

DAY 20

Fig. 7.2. Three-dimensional reconstruction of the third tiller bud of the same axis. Contours are 40 μm apart, and the outline trans-sections are for the levels indicated.

drawings for tiller 2 in Fig. 4.7.3. The series is intended to stress the tightness of the packing during tiller bud initiation.

An important point to keep in mind is that early tiller bud development bears very little resemblance to the development of a succession of leaf primordia within an apex. The bud grows in close contact with surrounding tissues, but the other condition for the prediction of exponential growth is not present – membership in a sequence of like

200

Triticum aestivum

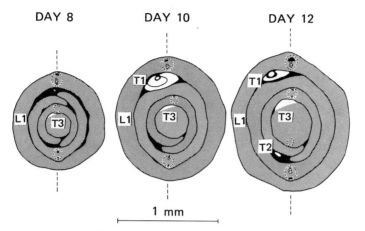

Fig. 7.3. Outline drawings of stem section at the level of tiller 3.

primordia. The more mature tissues of the axillant leaf sheath may be moulded outwards slightly by a vigorous tiller bud, but most of the accommodation for its growth is sustained by the softer tissues of the internode on the inside, thus yielding the flattened, wedge-shaped cavity of Fig. 7.1 (but see also Figs. 7.2 and 7.4). This, of course, is also the site of most protracted growth of the intercalary meristem of that internode, and it seems likely that the continued growth of the tiller is made possible by the enlargement of the cavity along with the growth and maturation of the internode. If this were all that could happen, no tillers would ever reach the light, and thereby achieve independence from the parent shoot. That some buds do 'escape' and grow vigorously prompted the detailed descriptive experiments to which reference has already been made (Williams *et al.* 1975; Williams and Langer, 1975).

The essentials of the experiment were that a wheat cultivar which rarely produces an effective coleoptile tiller was chosen, such tillers being variable in size and likely to have correlative effects on tiller 1. All plants were grown in long days so as to keep tillering down to manageable numbers, and two nitrogen levels each at two light intensities were the treatments used to produce differences in vigour. We can designate the treatments as L_1N_1, L_1N_2, L_2N_1 and L_2N_2, with the usual connotations. Under the conditions of the experiment coleoptile tillers grew slowly for several weeks but never exceeded 1 mm in length;

Triticum aestivum

TILLER 2

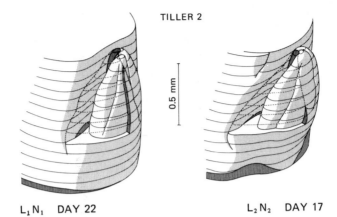

L₁N₁   DAY 22   L₂N₂   DAY 17

Fig. 7.4. Three-dimensional reconstructions for tiller 2 each with the same height but from the extreme treatments, L_1N_1 and L_2N_2 (see text). The near halves of the prophylls have been removed to give a clearer idea of the spatial relations within and behind the buds. The contours are 80 μm apart.

the low nitrogen treatments produced only three primary tillers, T1, T2 and T3, but about one plant in four receiving high nitrogen had a T4, though these never 'escaped'. In the other three plants T4 was completely suppressed.

Figure 7.4 depicts T2 at about the same stage of development for the extreme treatments, L_1N_1 and L_2N_2. Even so, that for L_2N_2 has 50 per cent more volume, and it achieved the same height five days earlier. The most obvious difference between the two buds is the visual evidence of greater outward thrust with L_2N_2, and we know from Fig. 7.5 that this tiller was just on the point of 'escaping', whereas that for L_1N_1 continued to grow in its cavity, and to five times its volume, before it too 'escaped' sixteen days later.

The growth curves of Fig. 7.5 and the relative growth rates of Fig. 7.6 can only be described as discontinuous, with the link at the time of 'escape'. There is, of course, no discontinuity in the growth curves of tillers which do not 'escape'. Growth is never exponential, and the form of the growth curve prior to 'escape' is best understood as a response to the ever increasing constraint of the surrounding tissues. Buds can 'escape' only when the thrust of growth exceeds the constraint, and this can happen at widely different times and stages of bud development, or not at all. Clearly, then, physical constraint is not the only causal

202

Triticum aestivum

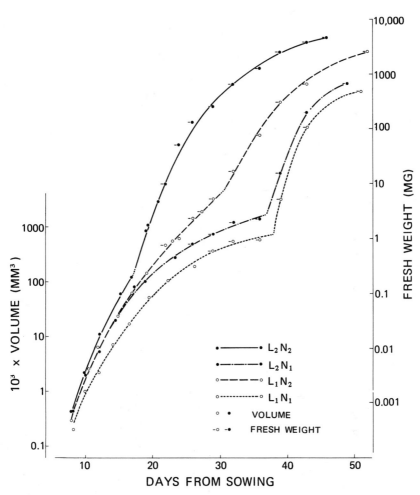

Fig. 7.5. Volumes and fresh weights for tiller 2 as these are affected by two light intensities and two levels of nitrogen supply. They are plotted on equivalent logarithmic scales, and the fresh weight scaly is extrapolated downwards to provide estimates based on an assumed constancy of weight per unit volume.

element in this sequence of events; growth, with all that that implies in metabolic activity and hormone action is a basic element, but constraint determines the *rates* of early tiller growth, and the interplay of opposing forces provides the trigger for tiller emergence and eventual independence from the parent shoot.

The desirability for experimental confirmation of this rather unortho-

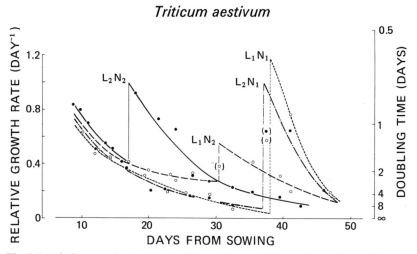

Fig. 7.6. Relative growth rates for the tillers of Fig. 7.5. The vertical sections of the curves relate to the discontinuities of the curves of Fig. 7.5, and the values in parentheses are based on intervals which include the discontinuities.

dox explanation will be obvious, and a recent experiment (Williams and Metcalf, 1975) is relevant. These workers submitted the same cultivar, 'Kalyan Sona', to two types of treatment when the first leaf was on the point of breaking through the coleoptile. Short sleeves of plastic tubing with internal diameters of 2.0, 1.7 and 1.5 mm were slipped over the coleoptiles and kept as close to the grain as possible. In other seedlings a small slit was cut in the coleoptile above the site of the coleoptile tiller. The experiment was properly replicated and harvests were taken at weekly intervals. Representative plants taken on day 21 for the control and three of the four treatments are shown in Fig. 7.7 and tiller lengths for this stage are given in Table 7.1. The wider tubings (2.0 and 1.7 mm) and the surgical treatment had no significant effects on the yields or linear dimensions of the primary shoots, but the narrowest tubing had a strangling effect, and in many of the seedlings of this treatment the coleoptile tiller forced its way through the base of the coleoptile below the tubing. This is seen as a quite secondary, though interesting phenomenon. Tillers 1 and 2 were suppressed by the sleeves, but grew normally in the controls and the surgically treated plants. Coleoptile tillers grew vigorously in a majority of the plants of the latter treatment, but did not emerge in the controls or in the plants with sleeves, except occasionally and belatedly by bursting through below the sleeves as was common in the treatment with the narrowest sleeves.

204

Fig. 7.7. 21-day wheat seedlings from an experiment designed to test the effects of mild constraint and release from contraint on the growth of tillers: C, control plant. $R_{2.0}$, Plant grown with a plastic sleeve having an internal diameter of 2 mm. $R_{1.5}$, The same, but with an internal diameter of 1.5 mm: S, plant whose coleoptile had been slit just above the site of the coleoptile tiller. ($\times 0.4$.)

Table 7.1. *Tiller length (mm) of 'Kalyan Sona' wheat as affected by the treatments of Fig. 7.7*

Treatment	Coleoptile tiller	Tiller 1	Tiller 2	Tiller 3
Control	4.0	*178*	*28*	2.8
$R_{2.0}$	3.0	7.7	4.3	2.6
$R_{1.5}$	*188*	5.8	2.8	2.2
S	*80**	*154*	*85*	3.4

* Includes one very large and one very small tiller.
Statistical analysis gave a very high interaction between treatment and tiller number. Values in italics are significantly greater than other values in the same column.

The simplest explanation of these results is in terms of physical constraint. Its removal in the surgical treatment permitted active growth by the coleoptile tiller, and the application of mild additional constraint greatly reduced the length growth of tiller 1 and tiller 2. Incidentally, such an explanation provides an attractive alternative to that advanced by James and Hutto (1972) for their effects of tiller separation on yield in *Lolium perenne*. Severing the connecting tissue at the base could scarcely do other than reduce constraints upon young tillers within the coleoptile, the prophyll of the second tiller and within the early leaf sheaths.

8 Plant growth as integration

'... a most influential part of the physiological environment of any single cell is provided by its neighbours ... one cell's total environment thus differs to a greater or lesser degree from that of any other.

The march of relative growth rate does not therefore simply reflect changes in external factors, but is determined very largely by those internal processes of organization which set limits to the growth of practically all parts.'

RICHARDS (1969)

This chapter attempts to draw the threads together, and will be rather more speculative in character than those that have gone before. Not only does the presentation of masses of descriptive matter require the cement of integrative speculation; the very emergence of a truly quantitative biology would seem to depend upon it. As pointed out in the introduction, it was found impractical to present the facts about growth and development of a range of shoot-apical systems (Chapter 4) without drawing attention to situations in which physical constraint seemed likely to be relevant to the interpretation of events. We can now take a closer look at those situations to see if they are consistent with the proposition that physical constraint is indeed a significant element in the genesis of form and the determination of rates of growth of various primordia. The proposition has already been developed by presenting the essential results of a study of the initiation and growth of tiller buds in wheat. This system is seen as a valuable test system for these ideas.

We will then look again at the organization of the shoot apex, and seek to discover if there is indeed anything more subtle in leaf arrangement than that which 'lies in the steady production of similar growing parts, similarly situated, at similar successive intervals of time'.

8.1 PHYSICAL CONSTRAINT AS A DETERMINANT OF GROWTH RATE

It was suggested for flax that the growth rate of the first pair of leaves may have been limited by that of the cotyledonary petioles, between which they arise (Figs. 4.1.1 and 4.1.15). It was also suggested that the growth rate of young primordia in the transverse plane is set by R_r, the radial relative growth rate of the apex; that growth in length, except in the first pair of leaves, seemed not to be under external constraint for the first 30 days; and that marginal growth was free to make a positive contribution some days after primordium initiation. After 30 days, a curious crowding of the leaf primordia occurred, and was believed to be responsible for the vagaries in the relative growth rate patterns for the upper leaves (Fig. 4.1.18). These interpretations rest quite heavily on the supposition that meristematic tissues, in these circumstances, are growing at rates below their potential. The brief period of rapid growth immediately after the initiation of each leaf primordium supports this contention (Figs. 4.1.17 and 4.1.18). Minimum doubling times of from 9 to 12 hours would then be achieved only momentarily, and Fig. 4.1.3 supplies visual evidence that such high rates could scarcely be maintained in the crowded conditions of the inner zone of primordia. The secondary peak in relative growth rate (Fig. 4.1.18), which occurs in the earlier leaves six to eight days later, is due to the activity of the marginal meristems (Fig. 4.1.1).

A rather more subtle example of possible effects of physical constraint is provided by Fig. 4.1.13 and the discussion relating to it. It is suggested that there may be quite violent gradients in axial growth rate; high in the dome itself, falling in the sub-apical region to rather low minima, then rising to maxima which may be higher than the rate in the dome. Figure 4.1.3 shows that the sub-apical region is one of intense morphogenetic activity: changes in form are extremly rapid, and there is plenty of scope for the interplay of mechanical stresses in the epidermal and hypodermal layers. Is it too fanciful to suppose that the low axial growth rate in this region is causally related to this interplay of stresses? Immediately below this region, change in form quickly becomes minimal, and there would seem to be relatively little physical impediment to axial growth, so long as the tissues remain sufficiently meristematic to permit continual cell division and active extension growth. This last remark is a salutary reminder that all maturing organs can quite properly be described as being subject to increasing physical constraint. After all, differentiation and maturation are mostly one-way processes, with general increases in cell-wall thickness and the establishment of rather rigid vascular systems. Continued growth is not compatible with these trends.

The early growth of the leaf of subterranean clover provides us with a new phenomenon – that of an extended period of strictly exponential growth (Figs. 4.5.8 and 4.5.9), but, before examining this in terms of physical constraint, it should be noted that the earlier leaves of the primary shoot exhibit high relative growth rates immediately after initiation in much the same way as was found for flax. This effect is no longer present after leaf 11, and Fig. 4.5.9 suggests that this is due to the increasing tightness of packing about the apex.

It has already been noted that, after the early formative stages of development, clover leaf primordia follow each other in what can be described as a developmental tunnel in which the primordia themselves make up an exponential series of similar forms (Fig. 4.5.1). If one member of the series were to grow relatively faster than its predecessor it would crumple up behind it, but if it grew more slowly a gap would open up between them, and the system would be an inefficient producer of leaves. Intuitively, then, one sees what actually happens as the optimal solution of a developmental problem. According to Rosen (1967), 'The optimal solution is the one which will meet all the conditions of the problem; i.e. will carry out the given task within the given context, with the minimum cost'. The difficulty seems to be to define and evaluate the appropriate cost factors. In the meantime the biologist must be content to seek generalizations and test their predictive value. One such generalization might be that a succession of similar leaf primordia produced at similar intervals of time and remaining in close contact will tend to grow exponentially. A necessary corollary is that such primordia will grow at a rate which is less than their potential rate of growth. As we shall see, there are already some grounds for believing that this generalization and its corollary do in fact possess predictive value. Figures 4.5.8 and 4.5.9 show that subterranean clover leaves fill the conditions and do in fact grow exponentially through more than three logarithmic cycles of size (from about 1 to 3000 units of volume). They escape from the constraint of the bud only when extension growth of the petiole carries the blade into the open air.

As already indicated, the shoot apex of wheat exhibits a special feature in that newly initiated primordia appear not to have any developmental space into which they may grow freely (Figs. 4.7.2 and 4.7.13). At a later stage, and when still highly meristematic, each leaf in turn is released from constraint by differential directional growth in its predecessor. The net result is that wheat leaf primordia grow exponentially for anything from a few days to a few weeks, according to leaf number, and then achieve, briefly, a relative rate of growth more commensurate with the potential for primordia which are still highly meristematic (Fig. 4.7.14). The chief difference between the clover and wheat systems lies in the fact that, in clover, maturation sets in concurrently with release from constraint and emergence from the bud, whereas in wheat maturation and emergence occur together some days after release from constraint.

This difference would seem to provide very strong evidence indeed in favour of the relevance of constraint as a determinant of growth rates in leaf primordia.

A phenomenon which has puzzled me for a long time is that several internodes in the sub-apical region of the stem are similar in length (Williams, 1960; also Figs. 4.7.11 and 4.7.12 above). This in itself implies a very low relative rate of growth, and the analysis of Chapter 4 (Table 4.7.2 and text) confirms this. As in the parallel case for flax, it is suggested that the sub-apical region is one of intense morphogenetic activity, and that the interplay of mechanical stresses may be sufficient to account for the near cessation of growth in an axial direction during the period of initiation and early growth of the primordia (Fig. 4.7.13).

Another observation on wheat which can be explained in terms of constraint is the difference between the growth curves for the lemma and the palea during early spikelet development. Only the palea shows an early lower rate followed by a rise, as for leaf primordia; and only the palea arises in close contact with neighbouring members of the spikelet (Figs. 6.1 and 6.6).

The other shoot-apical systems of Chapter 4 exhibit growth rate phenomena already reported for flax, clover, and wheat, and also a few novel features. High values immediately after primordium initiation are demonstrated or inferred for tobacco (Fig. 4.2.10), cauliflower (Fig. 4.3.5 and text), lupin (Fig. 4.4.3 and text) and *Eucalyptus* (Fig. 4.6.12 and text). In all cases the three-dimensional reconstructions show that the developmental space above the apices was free to accommodate the outcome of these temporarily elevated rates. This is also true for flax and the early primordia of clover. Only in wheat, and for the later clover leaves was such a developmental space not available. *Ficus* provides a special case as we shall see in a moment.

The existence of a period of exponential growth for tobacco leaf primordia (Fig. 4.2.10) is something of an embarrassment to our prediction concerning such growth, for Fig. 4.2.1 does not give the impression that successive primordia are in close contact near the apex.

However, they *are* tightly embedded in a mass of large hairs, and there seems no reason to doubt that a felt-like mass of hairs could just as readily impede the growth of young primordia as would the presence of closely packed, neighbouring primordia. Fig. 4.2.4 shows that the hairs do not surround the apex and the very young primordia, so, once again, developmental space is available for an initial peak in the relative growth rate of the primordium.

The growth rate data for cauliflower and lupin (Figs. 4.3.5 and 4.4.3) are indirect, but there seems not to be a strictly exponential phase, and the secondary peaks are best interpreted as evidence of the activity of marginal meristems producing the blade as a whole in cauliflower, and those of the seven folioles in lupin. In this they parallel the events for the earlier leaves of flax.

Although the data for *Eucalyptus* are not as precise as may be desired, they are consistent with the notion that physical constraint is a factor in determining the growth rates of leaf primordia. In both species examined, the closeness of packing of primordia in the bud increases markedly with time, and this is correlated with quite dramatic reductions in relative growth rate following the high initial values already noted.

The large terminal bud of *Ficus* was studied because its successive members are packed tightly together; and the prediction of exponential growth was amply justified. The leaf, without the stipular organ, grows exponentially for more than 40 days, and in so doing spans a range of six logarithmic cycles in size. The seeming anomaly that the stipular organ at first grows relatively faster than the leaf itself, and later settles down for about 20 days to an exponential rate which is lower than that of the leaf is readily understood when it is realized that the developmental space into which the stipule is growing is itself changing its form. It seems that an alternating succession of dissimilar forms growing in close contact is unlikely to produce exponential growth in both members. In this case the stipule was rapidly changing form at a time when there was little change in leaf form. Later, when *neither* structure was changing its form appreciably they *both* grew exponentially but at different rates. This latter paradox is explained in Chapter 4 and need not concern us here.

To sum up this section, it would appear that relative rates of growth on leaf primordia are likely to be highest when developmental space is freely available to them, and before the processes of maturation begin to limit their potential rates of growth.

When like primordia succeed each other at similar intervals of time, and they remain in close contact, they can be expected to grow exponentially at a rate which is lower than their potential rate, and presumably at that set by the oldest member still in the system.

It is suggested that escape from such systems may require a specific event, such as the rapid expansion of the petiole, as in clover.

These tentative conclusions presuppose that physical constraint is an important determinant of growth rates, possibly at many stages during development, and while direct experimental evidence supporting such a claim seems likely to be slow to accumulate, there is plenty of scope for the comparative study of plant biological systems within which physical constraint seems likely to play a part. Thus, further study of different types of shoot-apical systems is clearly necessary. We already know that the rate of root growth is sensitive to the physical constraints of the rooting medium (Greacen and Oh, 1972; Gill and Miller, 1956), but there is need for study of the physiology of this response. Other systems which could well yield interesting results are those relating to winter dormancy and subsequent bud burst in deciduous plants, and those of seed maturation, dormancy and subsequent germination.

8.2 CONSTRAINT AND THE GENESIS OF FORM

That physical constraint should play a part in the genesis of form is perhaps a more subtle notion than that of its effect upon growth rates in various primordia. Nevertheless it must often have occurred to developmental morphologists when they contemplated the very beautiful and efficient packing of parts within shoot-apical systems. Some examples are provided above for flax (Figs. 4.1.1, 4.1.2 and 4.1.7), tobacco (Fig. 4.2.3), *Brassica* (Fig. 4.3.2), clover (Fig. 4.5.4), *Ficus* (Fig. 4.8.4) and *Dianella* (Fig. 3.2). Indeed the whole array of patterns which is associated with phyllotaxis can scarcely be explained except in terms of mutual pressures, unless one is quite impervious to the claims of space-filling considerations as acceptable causal mechanisms. At the very least, the patterns must be accepted as optimal solutions to developmental problems. Adult form can of course be greatly modified by later events involving differential patterns of cell division and expansion. The abaxial positioning of the prominent venation, particularly the mid-veins of leaves coincides with the near-necessity of such positioning within the bud.

Some specific examples of the possible operation of constraint in the

212

genesis of form have been noted above. Thus it is suggested that flax leaves change from being linear lanceolate and rather thick to become more pointed, broader towards the base and distinctly thinner at the edges because of increasing pressures within the bud (Figs. 4.1.1, 4.1.2 and 4.1.6). R. Snow (1965) implies that spirodistichy in certain cucurbits may result from distortions due to precocious axillary buds, and a similar argument could be advanced for spirodistichy in subterranean clover (Fig. 4.5.4D). On another occasion Snow (1952) has suggested as an explanation of the curious phyllotaxis of *Costus cylindrica*, that each primordium arises where it does because the tilt of the apex throws one side of it against the flank of an earlier primordium, and that the pressure there is enough to retard leaf determination. There is a close parallel here with the situation in the sloping apex of *Ficus*, for which it has been shown that the apex comes into direct contact with the stipule of the youngest primordium, but not where the new primordium will arise (Figs. 4.8.2B, 4.8.3 and text). In this case the divergence angle is *not* deflected from the Fibonacci angle, even though the apex is almost completely encircled before the new primordium has initiated (Fig. 4.8.7). Concerning *Costus*, Snow is inclined to reject an earlier suggestion (Weisse, 1932), that the new primordium arises where the pressure between the apex and previous encircling leaf bases is least. Snow also quotes a very early suggestion by Nemec (1903) that pressure retards leaf formation.

The three-dimensional drawings for *Eucalyptus* (Figs. 4.6.5 and 4.6.6) provide a clear case of deformation by constraint, for the tips of the youngest leaf pairs have obviously been forced together by the marginal growth of the previous pair. In the mature apex, too, there is little doubt that form is moulded by pressure, because the angular shapes, particularly in the petioles, are retained to maturity (Fig. 4.6.8).

Finally there is the evidence from tiller-bud development. The external shape and especially that of the prophyll was discussed earlier. Indeed the fourth tillers of the high nitrogen treatments are so perfectly moulded into their related internodes that they are difficult to see when the leaf sheath is removed. It was also noted in Chapter 6 that Arber (1934) and Turpin (1819), much earlier, had invoked pressure as a formative agent in the genesis of spikelet parts. More recently Williams *et al.* (1975) drew attention to the mid-vein of the axillant leaf as a determinant of the eccentric positioning of the tiller bud; to the likelihood that pressures or tensions determine the side on which the bud will be initiated, and that the strong tendency for all buds to be on the same side of a plane through the mid-veins is a related phenomenon (Fig. 7.3). They also suggest that the manner of rolling of the leaf, the under–over position, may be in response to constraints at the time the leaf margins actually meet (see the third, fourth and sixth diagrams of Fig. 4.7.3; the fifth diagram provides an exception to this rule).

Since most of the examples cited here are highly speculative interpretations of formative events in development, yet another speculation can perhaps be permitted. It concerns some observations on the uni- and trifoliolate condition in fenugreek, *Trigonella foenum-graecum* L. published by Rijven (1968). In the mature seed the only primordium normally present is that of the first leaf, which is unifoliolate. This leaf is initiated rather early, but if embryos are explanted before the primordium is in evidence, and successfully grown on an agar medium, the first leaf primordium is found to be trifoliolate. Rijven also found a plant whose seeds contained a high percentage of embryos with two leaf primordia, and the first two leaves of the seedlings from such seeds proved to be unifoliolate; only primordia formed later, after germination, became trifoliolate.

These observations are taken to suggest that a switch exists to the uni- or trifoliate pathway of primordium development, and that the setting of the switch corresponds to physiological conditions in the intra- and extra-ovular situation. Rijven discusses a number of possible mechanisms, but notes in passing that explanted embryos exhibit growth changes such as increased expansion and the opening of the cotyledons. In the light of our present discussion it is suggested for the intra-ovular situation that the possibility that actual physical constraint may inhibit the initiation of the lateral folioles should be taken seriously. In the extra-ovular situation the opening of the cotyledons would remove that constraint. Another look at Fig. 4.5.2 will possibly help the reader to see the cogency of this argument.

8.3 ORGANIZATION OF THE SHOOT APEX

It seems appropriate at this point to try to evaluate the main findings of Chapter 4 as these relate to the dynamic events within the apex proper. The many three-dimensional drawings of a considerable range of apex types, and the new information about rates of change within them is perhaps sufficient justification for an attempt which cannot but be highly speculative. The apex of *Brassica oleracea* (Fig. 4.3.1) seems to exemplify all the main features of 'normal' apices, and the fact that

Brassica oleracea

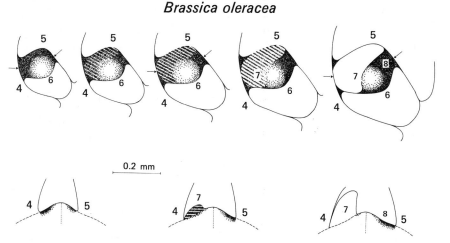

Fig. 8.1. Integrating diagram illustrating progressive changes in the vicinity of the shoot apex in *Brassica oleracea* during a single plastochrone. The upper series of drawings are in plan view and the first, third, and fifth are shown below in longitudinal section, in accordance with the arrows. The centre of the dome is indicated by the absence of stippling, and the gradient of intensity follows that of increasing meristematic activity.

it is rather flat and has relatively large primordia makes it a good choice for this exercise, especially as the distinction between activity in radial and axial directions seems all-important. Fig. 8.1 is a schematic interpretation of the progression of events during a single plastochrone in *Brassica*. The fifth diagram of the top row repeats all the essentials of the first, except that the orientation of the new bare apex has shifted by 137.5° and primordium 8 takes the place of 7 in marking the beginning of the plastochrone. The increasing hatching of the site of primordium 7 is intended to serve the double purpose of indicating its growth without altogether obscuring the postulated gradient of meristematic activity down the flanks of the apex. While it is undoubtedly true that the centre of gravity of the bare apex changes with each plastochrone, there seems no need to postulate any shift in the centre itself. Even with the large alternating primordia of *Pisum sativum*, Lyndon (1968) found no movement of his central zone. That there should be no such shift is in keeping with the continuing integrity of the shoot apex.

According to Clowes (1961, p. 60), most people working on shoot apices believe that all the cells above the youngest primordium are meristematic, but that there is a gradient of the sort indicated in Fig. 8.1. This impression is supported by the appearance of the cells, for those at the centre are more highly vacuolate and their nuclei tend to occupy

215

a smaller proportion of the cell volume. In our work with subterranean clover, too, we have noted that nucleoli are smallest at the centre, larger in the flanks of the apex, and larger still in very young primordia. Even if the evidence that summit cells divide less frequently than the cells on the flanks of the apex is rather weak, there seems little doubt that, in a vegetative apex, the central zone is one of relative quiescence, so that the periclinal divisions necessary to the initiation of a leaf primordium are likely to take place as far as possible from that zone. This is an essential element of the hypothesis of Snow and Snow (1931, 1947) that each primordium arises in the first space that becomes both wide enough and distant enough from the growing point. On this hypothesis primordium 7 of Fig. 8.1 must arise at nine o'clock relative to the centre, and the gap will be distant enough only when the vertical wall of primordium 4 has receded far enough to yield the necessary space between primordia 5 and 6. The drawings illustrate rather nicely why the wings of the new primordium, as seen in transverse section, are so dissimilar. Not only is the available space above primordium 5 greater than that above 6 (the younger), but the apex is eccentrically placed relative to the new primordium as a whole. The very shape of the inner margin of each primordium provides indirect evidence for the existence of the gradient in growth potential of which we have been speaking, and, to some extent, each new primordium seems capable of encroaching upon the apex itself until such time as the inner margin is well defined. This phenomenon is clearly shown in tobacco (Fig. 4.2.3), in *Eucalyptus* (Fig. 4.6.5) and in serradella (Fig. 4.9.1).

The foregoing conjectures concerning primordium shape can be restated in terms of physical constraint. Thus the new primordium is soon limited outwards by the quite formidable constraints of the established inner boundaries of primordia 4, 5 and 6, more or less in that order, and the progress of its inner boundary is slowed by the low potential for growth at the centre. One wing is stopped by primordium 6, and the other stops short of the site of primordium 8, presumably because periclinal divisions at that site have already been initiated by the thrust from below. Clearly these events will be accompanied by chemical gradients which necessarily enter into the causal sequence. Similarly, the phenomenon of encroachment of a new primordium upon the apical dome seems to demand localized stimulation of relatively quiescent cells to increase their rates of cell division. These considerations tend to strengthen my conviction that the two main theories of phyllotaxis – the field theory, and that of the first available

216

space – may be less antithetic than they appear to be. Their concepts and frameworks have to do with distinct levels in the organic hierarchy which is the organism.

Returning to Fig. 8.1 it will be evident that for plants such as cauliflower and tobacco, the new primordium (7 in the diagram) will, once properly established, begin to recede from the central area, and the process will begin over again. With elongated apices like those of wheat and clover, the new primordia encircle the apex, but there is room above them, sufficiently far from the tip of the apex, for the initiation of the next primordium (Figs. 4.5.2 and 4.7.2). The most difficult apex to fit into any generalized scheme is that of *Ficus* (Fig. 4.8.7), in which the bare apex is so 'residual', even at the end of a plastochrone, that it looks like a mere fragment of the flank of a rather large apex. It is certainly very difficult to visualize any credible basis for gradients in such an apex. How, too, is it possible for this apex to maintain a strict Fabonacci angle, unless it is because of the pattern of physical constraint set up by the immediately preceding, and very large primordia?

To this point we have been concentrating on surface views of the apex, so it is inevitable that events appear to be dominated by the emerging primordia. This impression is corrected by the longitudinal sections of the bottom row of Fig. 8.1. Cells at depth within the apex are rather large (Fig. 4.1.5 for flax, Fig. 4.6.7 for *Eucalyptus*, and Fig. 4.7.4 for wheat), but their continued division and their expansion result in growth which is mainly in an axial direction. The thrust set up by this activity at depth gains an outlet mainly through the shoot apex where resistance is least, and there seems little doubt that it provides the physical basis for the continuing stability and integrity of the apex. If one needs a visual prop for this notion, one only has to remember the shape and manner of growth of asparagus shoot tips, or of root tips, for that matter, when one ignores the complication of the root cap.

But what of the genesis of new apical systems? These normally arise in the axils of leaf primordia and, in the systems examined here, this happens in the sub-apical region of intense morphogenetic activity, at a very early stage. Hussey (1971) has shown for tomato that the whole of the apical surface is in an active state of cell division and expansion, except in the axillary region above the primordium. This is transformed into a passive region of non-dividing cells destined to produce the future axillary meristem. Hussey attributes these events to physical compression, and shows that micro-incisions made in the region close immediately. Perhaps the interplay of tension and pressure in the vicinity stimulates periclinal division at depth. This could explain the curious 'shell' patterns which are so often observed in longitudinal section as an early indication of bud initiation (Denne 1966*b*, for clover; Sharman 1945, for *Agropyron*; and in Fig. 4.6.7, herewith for *Eucalyptus*). Once this is established there is an easy progression to a normal shoot apex.

The growth potential, though slightly lower in the centre than on the flanks of the apex, gets rapidly less as one moves into the sub-apical region. This reversal of gradient may well underlie the curious papillose shape of the apex in *Brassica*, as in some other apices, and the gradient in general accounts for the radial symmetry of the apical dome (Fig. 3.1). However, the radial continuity of the system imposes exponential or near exponential growth in the transverse plane, but permits differential activity of a rather limited kind in an axial direction. In this connection, attention has already been drawn to the likely effect of the interplay of stresses in slowing down axial growth in the sub-apical region. One should not forget, however, that the genesis of the primordium itself is predominantly in an axial direction.

This brings us to what is perhaps the most novel feature of the findings of Chapter 4 – the existence of high, though transient relative growth rates of primordia immediately after initiation. And they are high only if, as in *Brassica*, there is space above the apex into which the primordia may grow freely. These high rates are thought to be confined to the products of the tunica layer of the presumptive primordium, and Hussey (1971) has shown, for the tomato primordium, that the increased rates result from increases both in cell division and cell elongation. The work of Lyndon (1968, 1970), however, suggests an alternative explanation which is set out in Fig. 8.2. Lyndon divided his pea apices, as seen in longitudinal section, into five regions, for which cell volumes and cell numbers were determined. He found little or no increase in the rate of cell division associated with leaf initiation and concluded that the formation of the leaf primordium and the subsequent growth of the apical dome resulted from changes in the direction of growth rather than changes in the rates of growth. I have reduced the regions to three in Fig. 8.2, and the stages (fractions of the ninth plastochrone) and ages apply to both parts of the diagram. The upper set of outlines is my understanding of Lyndon's interpretation of his results. The question at issue is whether there is in fact any kind of mass flow of cells into the primordium from the axis, or into the axis and then out of it again into the apical dome during the course of the plastochrone. If such movements are the rule within shoot apices, then it follows that

218

ORIGINAL INTERPRETATION

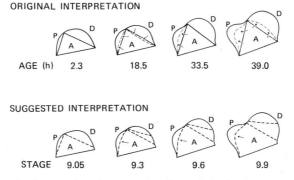

AGE (h)   2.3   18.5   33.5   39.0

SUGGESTED INTERPRETATION

STAGE   9.05   9.3   9.6   9.9

Fig. 8.2. Schematic diagrams based on a study of growth in the shoot apex of *Pisum* by Lyndon (1968, 1970). D, dome; P, primordium; A, axis. For explanation see text.

the interpretation of the high initial growth rates of primordia after initiation as a response to release from constraint is in doubt.

In a personal communication, Lyndon agrees that all his displace-ments of cells are artifacts in the sense that they depend on where the boundaries of the regions are placed. These are thought of as lines drawn between the leaf axils, *the only fixed points in the system*. This is, of course, a basic assumption in most of the new work reported in Chapter 4, and is quite credible for axils which have passed the formative stage. I doubt if they *are* fixed and identifiable in Lyndon's material (cf. the evidence for encroachment onto the dome by initiating primordia, especially for tobacco), except in the sense that one can identify the point of inflection in a given preparation. The critical point in the upper diagrams of Fig. 8.2 is the inflection between dome and new primor-dium. This could move up the flank of the dome and then back again in response to local differentials in growth and the accompanying stresses. Even the eighth axil may not have stabilized by the beginning of the ninth plastochrone. The suggested interpretation of Fig. 8.2 admits the possibility of some movement into the primordium, since this appears to be in the one direction throughout the plastochrone. It *may* be of consequence in large primordia but I doubt its significance in small ones. I am well aware that my own acceptance of a rigid distinction between tunica and corpus for purposes of volume estimation, and my extra-polations from primordium attachment areas is very much open to challenge, too.

In a more recent study of the pea apex, Hussey (1972) provides direct evidence for increased rates of cell division during primordium formation, and questions the validity of Lyndon's analysis in terms of subjectively determined regions. Even more important is his claim to have found evidence in Lyndon's data, as in his own, of the existence of growth centres in the corpus, similar to those previously reported for tomato (Hussey, 1971). These growth centres are local concentrations of cell divisions which precede primordium bulge formation by at least a plastochrone. Hussey suggests that the activity of such growth centres, may exert pressure on the tunica layers of the apical dome, thus accounting for the fact that the tunica layers pull apart when cut.

This discussion of organization within the shoot apex would be very incomplete without further reference to the rival theories of phyllotaxis. Fortunately this subject has been very thoroughly reviewed by Cutter (1965) and more incidentally by Richards (1948, 1951 and 1956), R. Snow (1955) and Wardlaw (1952, 1965). I have indicated my growing conviction that the field theory and that of the first available space may not be as antithetic as they appear. The third theory, that of Plantefol (1948) seems, as far as I understand it, to be purely descriptive in character. Concerning the two main contenders, Cutter (1965) also feels that they are becoming less opposed than formerly. However she qualifies this with, 'a minimum free space is certainly an essential prerequisite for the determination of a leaf site, but it is a sustaining condition, not an active factor; moreover, its creation probably depends on physiological events in the apex...'. This assessment undoubtedly reflects a widespread opinion, but it has the unfortunate effect of making a poor relation out of the morphologically oriented theory. Why do we so readily accept the view that physiological events are somehow more meaningful than morphological ones? Rather should we recognize such value judgements as expressions of the false antithesis between structure and function (Woodger, 1929, p. 326) and see that developmental morphology implies an interplay of forces. Gradients of growth potential within the shoot apex clearly supply positive thrusts which are matched by physical constraints within a dynamic whole.

I suggest, therefore, that the field theory should be modified to include mechanical as well as chemical gradients. If this were done it would be compatible with much of the observational and experimental evidence which is now seen to support the available space theory, and an inadequate title could be dropped. To distinguish it from the mutual repulsion field theory, the modified theory might be styled the mechanicochemical field theory. This would cover a sequence of events from the establishment of growth centres in the corpus tissue of the sub-apical region, through the initiation of periclinical divisions in the tunica, to the establishment of primordia under the mutual influences of older leaf primordia. In fact, it is a cycle of events.

Various experiments with growth hormones and inhibitors have failed to yield permanent or regular modifications of the phyllotactic systems of treated plants (Schwabe, 1971). In the main they have yielded only abnormalities of various kinds. Schwabe, himself, reports a number of experiments which achieved some success in this direction.

Using tri-iodobenzoic acid, Schwabe induced a change in the phyllotaxis of chrysanthemum from a low-order Fibonacci spiral to a simple distichous arrangement. Inevitably the divergence angle was greatly increased, but there was also a marked change in the shape of the apical cone, and with it an increase in the plastochrone ratio and apparently a lengthening of the plastochrone itself. These changes imply drastic changes in the space relations in the zone of primordium initiation, and Schwabe's drawings of transverse sections of control and treated plants suggest that the stipular wings of the young primordia could well stabilize the treated apices in the distichous condition (cf. the clover apex of Fig. 4.5.4, and that of *Dianella* in Fig. 3.2). Schwabe concluded that his data would support only a mutual repulsion field theory, but I see no hindrance to the building of an explanatory schema which recognizes the relevance of mechanical as well as chemical correlates. One has in mind the interpretations put forward by Snow and Snow (1935) of their surgical experiments with *Epilobium*. These authors split the decussate apices with vertical cuts in one of the diagonal planes and, of those that regenerated strongly, nearly 75 per cent gave spiral phyllotaxis.

In concluding this chapter on plant growth as integration, it may reasonably be claimed that a case has been made for acknowledging physical constraint as an important element in the processes of plant growth and development. This has been done by the critical use of the concept of relative growth rate, and the precise quantitative description of shoot apical systems. Physical constraint should not be thought of solely in its negative aspect, rather is it an essential element in the processes of organization which also include cell division and expansion. I see the notion of constraint as especially relevant to the generation of pattern, and the shoot apex is the place *par excellence* for its development. There is, of course, no suggestion that co-ordination by biochemical means is not also important. We are dealing with one of the interfaces between levels in an organic hierarchy and events at the higher level will necessarily be mediated by change in the lower, and *vice versa*.

Plant growth as integration

Perhaps the most pressing need for the near future is to extend our knowledge of growth and form in shoot-apical systems. Comparative, quantitative morphology of these objects will certainly yield valuable insights in its own right; it will also bring to notice test systems which will be amenable to experimental attack. One such system – the tiller buds in grasses – has already been examined in a preliminary way in Chapter 7. Experimental attack on the problems of leaf arrangement is still very much in its infancy, but it is hoped that this book will have helped just a little to remove the mystery and magic which has tended to envelop the subject of phyllotaxis.

Appendix

This appendix is primarily concerned with procedures which were developed for the quantitative studies of shoot-apical systems. It includes some suggestions on data processing, but it cannot be said too often that further studies of the kind will demand flexibility of approach and an awareness of the necessity for precision.

The final item is a set of tables for the conversion of relative growth rates (R day^{-1}) to doubling times.

A.1 THREE-DIMENSIONAL RECONSTRUCTION

The three-dimensional drawings which are used throughout this book are based on serial sections of the structures they represent. They have been very helpful in portraying changes in form and in identifying critical events in the genesis of form. One of my colleagues has labelled them the Michelin-type drawings, for rather obvious reasons, and others have thought the contour lines were too prominent. However, the lines do contribute greatly to one's awareness of form, as will be seen in Fig. A.2E and F. The procedure has been published in detail (Williams, 1970), but the essentials are repeated here for convenience.

Fig. A.1 illustrates the principles by reference to a sphere of radius 34 μm. Let us suppose that this sphere has been serially sectioned by horizontal planes 10 μm apart, and such that one plane passes through the centre of the sphere. The contour lines of the sphere are shown in plan view in Fig. A.1A superimposed on a square grid whose individual members have sides of 20 μm. The three lower contours are, of course, identical with the three upper ones, and all lie within the equatorial contour.

In Fig. A.1B, the same grid is shown foreshortened so that its members are 20 μm by 10 μm, and below it is drawn a vertical scale with 10 μm intervals (the section thickness), numbered serially from below upwards. Over this grid and scale is placed a piece of plastic tracing film which can be positioned by a pencilled axis (not shown) above the scale and as indicated by the upper arrow. The tracing film is required to be moved so that its lower arrow is successively opposite the scale divisions 1, 2, 3, etc. With this arrow pointing to 1, with intercepts proportional to those of its circle (Fig. A.1A) is drawn in. The result is an ellipse whose minor axis is half its major axis (Fig. A.1B). The lower arrow is then moved to scale division 2, and the process is repeated for contour no. 2, giving a larger ellipse 'below' the first (Fig. A.1C). Repetition for contours 3, 4 (Fig. A.1D), 5, 6 and 7 produces the full set of ellipses of Figure A.1E. It is not necessary to extend any contour beyond an intersection with the one immediately 'above' it, for that part of it

223

Appendix

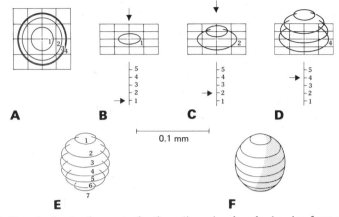

Fig. A.1. Steps in the development of a three-dimensional scale drawing from a contour plan. (After Williams, 1970.)

will be out of sight in the finished drawing. When the contours are finished, the 'edges' of the solid form are drawn in. These will occur wherever it is possible to draw common tangents to two or more adjacent contour lines. The finished drawing (Fig. A.1F), though giving a satisfactory three-dimensional effect, is slightly distorted.

Turning now to a natural object, the shoot apex of *Linum usitatissimum* L., the essentials of the procedure are illustrated in Fig. A.2. We have usually cut our sections at 10 μm, and stained them with a safranin-orange G-tannic acid combination (Sharman, 1943) which stains meristematic walls blue-black, thus giving distinct outlines. Some form of projection microscope is almost essential, though we have sometimes worked from photomicrographs. We have used a Reichert Lanameter (now styled the Reichert Visopan projection microscope), which eliminates the need to work in a darkened room. The camera lucida is not recommended.

Outline drawings of required sections are traced in pencil on thin tracing film (Figs. A.2A–D). Up to about six tracings can be superimposed on one film if a satisfactory reference axis can be established and the images 'move' in a systematic way. Coloured pencils or differing line types can be a useful aid here (Fig. A.2A). Fig. A.2 illustrates the procedure in detail for one leaf primordium (L19) only. Six consecutive outlines are given in Fig. A.2A, four alternate sections in A.2B, and two in each of A.2C and A.2D. The sections are numbered basipetally. The establishment of the reference axis represented in Figs. A.2A–E by the star and diagonal line – is sometimes difficult, and usually demands a reasonable visual knowledge of the structures being drawn. These drawings for L19 were not made in isolation, but along with those for other leaf primordia and the apical dome and stem (Fig. A.2F). Below the level of the dome there were many elements of symmetry to guide the super-position of images, but above it one had to rely on surrounding structures not included in the finished drawing. Experience will suggest other aids to this

224

Linum usitatissimum

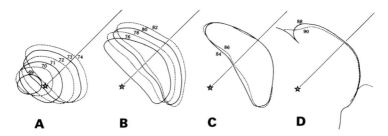

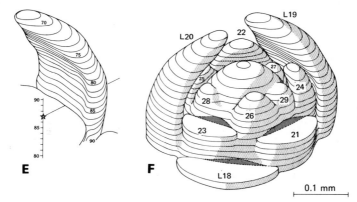

Fig. A.2. The development of a three-dimensional scale drawing of a leaf primordium and the shoot apex of an 11-day seedling of flax, *Linum usitatissimum*. For explanation see text. (After Williams, 1970.)

process, and adjustments can even be made in the composite drawing if need be.

The next step consists in translating the tracings into the perspective form of Fig. A.2E. The tracings are placed over a 16 cm by 16 cm grid, the star over its centre and the diagonal over the receding diagonal of the grid. The final drawing is developed on tracing film over a 16 cm by 16 cm foreshortened grid (cf. Fig. A.1B), with a scale covering the full range of section numbers to be transferred. The essentials have been described above for the sphere of Fig. A.1. The odd-numbered contours not given in Figs. A.2B–D are interpolated in A.2E. Even without the shading, Fig. A.2E conveys a great deal of information about the form of a primordium which is still only 0.2 mm in length. Fig. A.2F is the finished drawing for the whole apex, and shows the dome and 12 primordia (L18–29). Three of the larger primordia are dissected away, leaving their buttresses only, and permitting a clear view of the apical dome and small primordia.

A special problem, so far ignored, arises when sections are thick relative to the scale of the drawing, and some of the edges of the irregular frusta which

Appendix

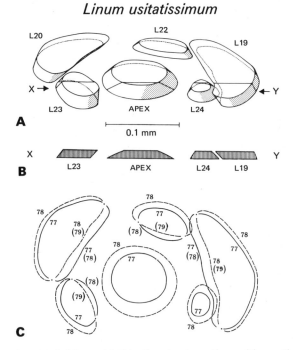

Fig. A.3. Procedure needed to avoid distortion due to sections with receding edges. In C, outlines appropriate to section 77 are shown as continuous lines, those for 78 as broken lines. Figures in parentheses indicate the derivation of hidden parts of these outlines. (After Williams, 1970.)

make up a given section have negative slopes. This situation is illustrated in Fig. A.3A, where the frusta of section no. 78 (Fig. A.2F) are shown in perspective, though viewed from a slightly different angle. The system is also shown in vertical section, XY in Fig. A.3B. The convention we have adopted is that the outlines required for a given section shall be those of the lower faces of the frusta that make up the section. Thus, when seen from above in the microscope, the frustum of the apical dome makes no problem, for its thin outer edge is the correct one (marked 78 in Fig. A.3C). The thin outer edge of the frustum of L24 is also correct, though part of it is obscured by the overhang of L19. The thin outer edges of the abaxial faces of the other primordia are likewise correct, but the lower, adaxial edges are obscured by the overhang. However, they are readily derived from the thin adaxial edges of the section immediately below (marked 79 in Fig. A.3C). Thus Fig. A.3C supplies complete sets of outlines appropriate to sections 77 (continuous lines) and 78 (broken lines). They are built up from the edges of sections 77 and 78, and 78 and 79 respectively as shown in the figure. Attention to this refinement can be rather confusing in practice, but is quite rewarding. Its neglect can yield strange distortions of form, and most structures are made to appear

226

thicker than is actually the case. Fortunately the occurrence of overhanging shapes is relatively rare in apical systems, and for larger structures drawn to a smaller scale the resulting distortions are negligible.

In spite of the slight distortion already mentioned, the projection has many advantages for depicting irregular solids made up, in the main, of continuous curved surfaces. It may be thought that the procedure described here is an inadequate substitute for good photography. Some very beautiful photographs of shoot apices do exist, but anyone who has tried to produce a satisfying developmental series of such objects will know how time-consuming and frustrating such attempts can be. Many apical systems, too, are covered by masses of hairs; these can be ignored by our procedure.

Since the usefulness of the procedure will depend to some extent on uniformity of presentation, a few practical suggestions are offered for inking in the drawings. The tracing film should be of good quality – we use one which is 0.004 inch thick – and should take ink well. We have worked mainly at magnifications of × 500 and × 130 and line thicknesses have been 0.2 mm for the contours and 0.3 mm for the edges. The shadow effect is produced with self-adhesive shading film, having a 10 % stipple and 27.5 lines to the inch. When applying this we visualize a light source on the left at a low angle with the horizon. The angle tends to accentuate the high spots – tips of leaf primordia, etc. – and gives a less artificial effect than horizontal lighting. Horizontal cut surfaces are left white, but vertical cut surfaces are shaded with a fine vertical hatching (also 27.5 lines to the inch) irrespective of lighting. The figures as published are much reduced – to less than one-third in some cases.

A.2 VOLUME ESTIMATION BY SERIAL RECONSTRUCTION

The procedure recommended by Wilhelm His (1888) has been used extensively throughout this book, and the importance that he attached to it has been mentioned in Chapter 1. The example of Fig. A.4 (see also Fig. 4.7.4) is for an 11-day wheat apex, and consists of a volume-distribution diagram and outline drawings of representative transverse sections. Those for leaves 4 and 5 are to the right of the diagram, and others for the apex and leaves 6 and 7 are to the left. The sections are numbered acropetally. Each leaf primordium can be regarded as being associated with a disc of insertion whose axial limits are taken as those sections in which the primordia appear half united with the axis (e.g. sections 11, 27 and 33 of Fig. A.4). The stem areas at these points are taken as those of the first complete stem section immediately above. Since the leaf primordia in wheat have been shown by Barnard (1955) to arise entirely by periclinal division of tunica cells (cf. Fig. 4.7.3), and the thickness of the 2-layered tunica is remarkably constant, these facts are used to define the boundary between the tunica, T, and the corpus, C, in Fig. A.4. The volume-distribution diagram is, of course, built up by plotting all individual areas appropriately against section number. The partial areas of the diagram are proportional to the volumes of the structures they represent.

While the building of such diagrams is advisable for each new system being estimated, it is possible, with experience, to measure areas directly on the

Appendix

Triticum aestivum

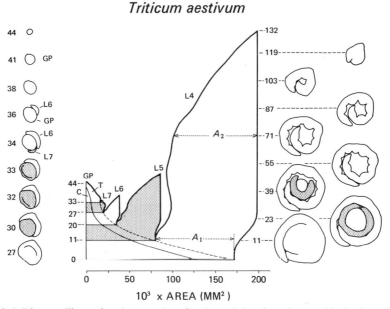

Fig. A.4. Diagram illustrating the procedure for determining the volumes of leaf primordia and related tissues of the shoot apex in wheat. For full explanation see text. GP, growing point; T, tunica; C, corpus; L4, L5, etc., successive leaf primordia. Based on an 11-day apex, 1.32 mm long. (After Williams, 1960.) For A_1 and A_2 see p. 230.

screen of the projection microscope, using a calibrated grid, and to plot only the problem areas involving boundaries. For direct integration, the areas of from 10 to 16 equally spaced trans-sections should be measured, summed and multiplied by the distance between sections.

Fig. A.5 and Table A.1 illustrates a variant of the procedure for a 15-day flax apex. In this case all the sections for each of the eight small primordia were traced, including the buttress bulges, and total areas determined for each primordium. The areas given in Fig. A.5 are those of the original tracings, and alternative values are converted to volumes in Table A.1. The stippled areas derive from section 506, which is shown complete. Except where the primordium is represented only by a buttress, each has a section, *B*, which represents its half junction with the axis. Indeed these are equivalent to the transverse projections which make up the phyllotaxis diagrams of Fig. 4.1.7, and their areas are used indirectly to estimate the tunica contribution to the volumes of the small primordia. Table A.1 shows that this tunica contribution becomes relatively larger the smaller the primordium, but so do the errors in the estimation of A_B, the basal area. However, the areas were found to form exponential series, so it was a simple matter to obtain improved estimates based on linear regressions of their logarithms. Table A.1 includes some corrected basal areas as well as two extrapolations (for primordia 34 and 36) which permit estimates of growth rates on the flanks of the apical dome.

228

Linum usitatissimum

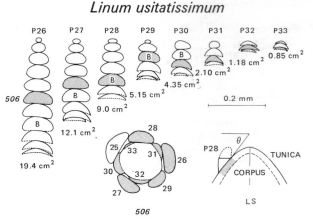

Fig. A.5. Determination of the volumes of small primordia in a 15-day flax apex, including the buttress and tunica components. The areas under P26 are from section nos. 501–511, etc. B marks the basal sections of the five older primordia. The longitudinal section is built up from residual stem areas such as the white area of section 506, and the angle θ is developed graphically as a function of primordium number: LS, longitudinal section.

Table A.1. *Volumes of small flax primordia (Fig. A.5)*

	Tunica				Blade and buttress		
Primordium	Basal area, (A_B) (cm²)	Sec θ	Attach-ment area (cm²)	$10^3 \times$ tunica volume* (mm³)	Area (cm²)	$10^3 \times$ volume** (mm³)	Total $10^3 \times$ volume (mm³)
26	2.75	1.53	4.21	0.123	19.4	0.285	0.408
28	1.80	1.48	2.66	0.078	9.0	0.132	0.210
30	1.17	1.43	1.67	0.049	4.35	0.064	0.113
32	0.78	1.38	1.08	0.032	1.18	0.017	0.049
34	0.51	1.33	0.68	0.020	—	—	0.020
36	0.34	1.26	0.43	0.013	—	—	0.013

  * The conversion factor ($\times$ 0.0293) covers magnification and a tunica thickness of 20 μm.
** The conversion factor ($\times$ 0.01467) covers magnification and section thickness (10 μm).

The tunica volume V_T is given by:

$$V_T = KA_B \sec \theta,$$

where K is the tunica thickness – 20 μm for our cultivar of flax – A_B is the basal area as before, and θ is the slope of the attachment area of the primordium (Fig. A.5). This angle is estimated for each primordium from the reconstructed longitudinal section. The essentials of the calculations are set out in Table A.1, which includes partial and total volumes for the primordia indicated.

Appendix

This procedure is essentially that adopted by Hannam (1968) for her study of the young tobacco plant. Both procedures are tedious, and each new application tends to produce its own minor problems. Precision is especially important when working at $\times 500$, and it is necessary to establish boundary conditions between primordium and parent tissue with care, and to apply them objectively on successive occasions. Boundary conditions are very much more important for early than for late stages in the development of a given primordium.

For the larger primordia, and especially if there are large numbers of similar ones to be estimated, one can set up regressions based on a minimal number of area and length measurements. For example, the following equation was used for wheat leaf primordia:

$$y = -0.3066 + 0.8646x + 0.0407x^2,$$

where y is the logarithm of the actual volume (including buttress and tunica component) and x is the logarithm of the estimated volume V', where

$$V' = L(A_1 + A_2),$$

and L is the total length, and A_1 and A_2 are as for L4 in Fig. A.4.

For the volumes of the larger flax primordia, we have used areas at four standard positions – 1/8, 3/8, 5/8 and 7/8 of the distance from tip to half-junction, and a correction factor for the buttress and tunica component.

A rather special problem was posed by the apices of *Eucalyptus*, for which the determination of the boundaries between primordia and stem were rendered especially difficult by the modified decussate system in *E. grandis*. For both species a solution was based on the rectangular shapes of the stem sections (Figs. 4.6.3, 4.6.4 and 4.6.10). Areas at the points of half junction of the primordia were converted by an arithmetic routine involving length–breadth ratios (Fig. 4.6.10) to the contact areas which are stippled in Fig. 4.6.4D. Once again, tunica thickness was remarkably constant (Fig. 4.6.7), so the tunica components of volume were easily determined and added to those for the blades and buttresses of the pairs of leaf primordia. Although we did not do this, stem internode volumes could have been estimated for the truncated solids of Figs. 4.6.3 and 4.6.4.

Perhaps enough has been said to indicate the need for flexibility in developing procedures for volume determination, and the need to adhere to the procedures quite objectively.

A.3 PHYLLOTAXIS

The system devized by Richards (1951) for the description of phyllotaxis has been presented in some detail at the beginning of Chapter 3, but there remain some practical considerations, particularly in relation to the estimation of the plastochrone ratio in spiral systems.

The best way to define the centres of most spiral systems is by trial and error after the preparation of transverse projections from sets of serial sections.

Linum usitatissimum

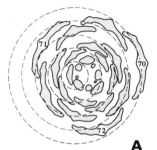

A

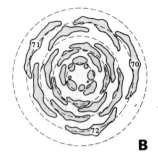

B

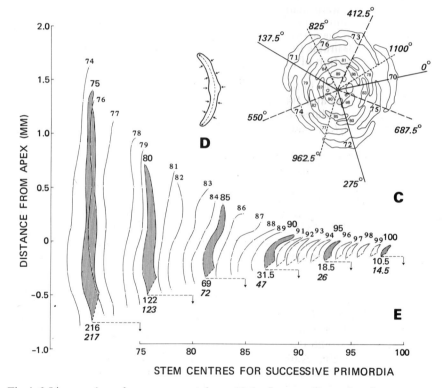

Fig. A.6. Linear and angular measurements from a 32-day flax apex. For explanation see text.

However, before doing this, we will examine some of the problems associated with the extraction of linear and angular measurements from a set of serial sections. The drawings of Fig. A.6 are for a 32-day flax apex. While it is difficult to obtain sets which are perfectly cut, any which are obviously skew cut should be discarded. This is most readily judged by the pattern of junction of successive primordia with the axis.

Appendix

In transverse section it is usually possible to centre the system with a set of concentric circles placed over the image, as in Fig. A.6A and B. This process is still easier when the apex or stem is in the section. Radial distances can be obtained directly or from tracings, and can be used to reconstitute the length profiles of successive primordia, as at E in Fig. A.6. True lengths are readily extracted from these profiles and, for every fifth primordium of the example, they are compared with estimates based on numbers of 10 μm sections from tip to base. While the estimates differ by less than 1 per cent for primordia 75 and 80, the discrepancy is almost 50 per cent for primordium 90, which curves over the apical dome.

Fig. 2.2 is based on these corrected lengths, and width and thickness measurements were made as shown in Fig. A.6D. The section chosen for each primordium is at a point 5/8 of the distance from its tip, the width is that of arc between the arrows, and the thickness is a mean for the five positions.

It is often desirable to have measurements, or simply to know the shapes in longitudinal section of primordia or whole apices which have been sectioned transversely. The procedure outlined above can achieve these things – admittedly at a cost in time and patience. Thus, median sections of primordia 75, 80, 85, 90, 95 and 100 are included in Fig. A.6E; the angle θ of Fig. A.5 is derived by a related procedure; and the median longitudinal sections of Figs. 4.5.10, 4.7.11 and 4.7.12 are all constructed from sets of transverse sections.

The measurement of divergence angles from transverse projections presents little difficulty once one has established the centres of the primordia and of the system itself. Indeed a single transverse section in the vicinity of the apex will give quite accurate results. This will be clear from Fig. A.6C in which the concentric circles of B are replaced by a summation sequence of eight Fibonacci angles – 137.5°, 275°, 412.5°, . . . , which can be drawn on tracing film. These show reasonable agreement with the placement of primordia 70–78 inclusive. For actual measurement one uses a transparent, full-circle protractor.

When it comes to the estimation of plastochrone ratios it is not wise to rely on single transverse sections, because radial distances in such sections do not necessarily reflect the transverse growth of the apex. We did so for *Mentha piperita* (Fig. 4.6.11), but only after making sure, from longitudinal sections, that the result would have been much the same if a transverse projection had been available. Fig. A.6E shows that young primordia of flax tend to grow outwards at the base and then inwards to varying extents. Clearly, the only sound basis for the estimation of plastochrone ratios is the transverse projection, and, again, each new apex tends to supply its own special problems. Fig. A.7 offers some suggestions on procedure.

For *Brassica* the apical cone is large and relatively flat, and the object is to establish the junction lines between primordia and cone, and between neighbouring primordia. In Fig. A.7 the primordia are indicated by the larger numbers; the small numbers relate to section numbers below the tip of the dome. The ninth and youngest primordium had only just initiated, and its rather hazy inner margin was derived from section 1. The inner boundary of

232

Brassica oleracea   **Linum usitatissimum**

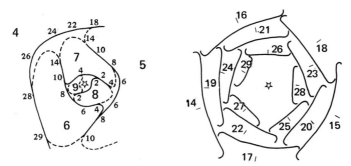

Fig. A.7. The development of transverse projections for cauliflower and flax. For explanation see text.

the eighth primordium was defined in section 2, but had disappeared in section 3. The inner boundary of the seventh primordium began in section 2, but most of it was defined in section 4; that of the sixth primordium began in section 4 and ended in 14; that of the fifth began at its centre in section 6, but fell away in both directions to sections 10 and 14 respectively; and the fourth began in section 18 at one end and fell all the way to section 29 at the other.

One's first reaction to the seeming lack of pattern in these inner boundaries is that the section must have been skew cut. However, because of the relative flatness of the cone, the slopes of the boundary lines would be quite small. Then, too, since the outer boundaries of young primordia are largely defined by the inner boundaries of their older neighbours, one can concentrate on defining the succession of inner boundaries. The importance of careful centring of each successive section during this process can scarcely be over-stressed. The less certain boundaries for this cauliflower apex are indicated by broken lines, and further steps in the establishment of the phyllotactic parameters are defined and discussed in Fig. 3.6 and the accompanying text. However, we will return in a moment to the establishment of plastochrone ratios for the two systems of Fig. A.7.

The development of the transverse projection for flax (Fig. A.7) was somewhat easier because of the many reference points for centring the sections, and because the inner boundaries of the primordia tend to be defined within a single section. The inner boundaries of primordia 29 and 28 of the example were in section 5, those of primordia 27 and 26 were in section 6; that of primordium 25 was in section 7 and so on. The lateral extent of the primordia can be taken from the projected images, and also their centres where the mid-vascular bundle is clearly defined. The outer boundaries can then be filled in by eye to yield the finished diagram (Fig. 3.1). We have measured radial distances from the centres of the inner boundaries, though the 'centres' of the primordia would have done equally well for determining plastochrone ratios.

The requirement that the plastochrone ratio should be determined for the region of primordium initiation (Fig. 3.6 and text) raises some problems, if

Brassica oleracea

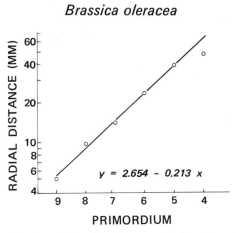

Fig. A.8. The determination of the plastochrone ratio from the radial distances of the primordia in a 22-day cauliflower apex. In the linear regression, *y* is the logarithm of the radial distance.

only because errors of measurement rise steeply the closer one gets to a centre which is itself an estimate. For the *Brassica* example a second attempt to locate the centre yielded the sequence of radial distances plotted on a logarithmic scale in Fig. A.8. The tedium is taken out of this exercise if the measurements are plotted directly on semi-log paper. A slight improvement in the relation of Fig. A.8 might have been achieved by moving the centre fractionally away from 9 and nearer to 8, but the regression line would have remained virtually as it is. The rejection of the value for 4 is based on the knowledge that any exponential relation which exists can be expected to break down eventually in the sense indicated by 4.

By definition, the plastochrone ratio is that for successive radial distances, and values based on primordia 8 and 9, 7 and 8, and 6 and 7 are 1.885, 1.429 and 1.679 respectively. These are too variable to be useful, so one must look for a more efficient procedure. Where, as in Fig. A.8, the radial distance relation is an exponential one, the best procedure is to use the coefficient of *x* in the equation. Ignoring the negative sign, this is 0.213 and its antilog, 1.633 is the mean plastochrone ratio between primordia 5 and 9.

In some circumstances a satisfactory estimate is obtainable from the radial distances of primordia which differ by three plastochrones. In Fibonacci systems, such primordia are on radii with a small angle between them, so that some major sources of error are eliminated. The plastochrone ratio is given by the cube root of the ratio of the measurements. It will be clear that one must not rely on single determinations. Mean values based on several apices of a given age should be aimed at.

The basic data for three flax apices of different ages are plotted in Fig. A.9. For day 18 the radial distances for all primordia from 16 to 43 are given, but for the other two the relation is built mainly on groups of three at intervals.

Linum usitatissimum

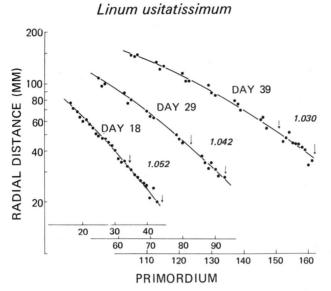

Fig. A.9. The determination of the plastochrone ratio from the radial distances of the primordia in flax apices of three ages. The estimates are given in italics.

It would be rather hazardous to base estimates of plastochrone ratio solely on the radial distances of the youngest primordia of these apices, though one knows that the relation must be very close to an exponential one near the apex. The curves were fitted visually by superimposing a graded set of curves drawn on tracing film, and reading off enough values to draw in the selected curve with a flexible ruler. One could, of course, fit quadratic regressions, but this was thought not to be necessary. Interpolations were then read off the graph for the most recently initiated primordium and the tenth one below it. For day 18 these were primordia 44 and 34 and their radial distances were 19.3 and 32.1 units respectively. The plastochrone ratio, r, is given by

$$\log_{10} r = \left(\frac{\log_{10} 32.1}{19.3}\right)\bigg/ 10 = 0.02209$$

and $r = 1.052$. Similarly for days 29 and 39 the estimates were derived from interpolations for primordia 93 and 83, and 161 and 151 respectively (the arrows in Fig. A.9). The plastochrone ratios for the two apices were 1.042 and 1.030 respectively.

The rather special case of *Eucalyptus*, and possibly other decussate systems, is discussed fully in Chapter 4, §6, where it is shown that one must not go beyond the first two leaf pairs to calculate plastochrone ratios. This is because precocious elongation near the apex permits a rapid falling away of the radial relative growth rate.

Appendix

A.4 AGE EQUIVALENCE AND COVARIANCE

At various points in Chapters 3, 4 and 5, the reader will have become aware of the close attention which has been given to the attainment of precision in the quantitative description of the growth of the members of shoot-apical systems, and of the chemical components of leaf primordia. Of the procedures which are there used, perhaps the most sophisticated is that of age equivalence, though this in turn is really based on the principles of covariance analysis. Surprisingly little attention has been given to these matters in the literature, so a brief account may be helpful.

The conventional procedure for determining a series of points on a growth curve for a plant growing under a specific set of conditions, is to make a succession of harvests of random samples, determine the means, and from these to estimate the growth increments and other derivatives in which the investigator may be interested. In many cases, random sampling of relatively large numbers of plants (as for Figs. 2.6–10 above) will give all the precision that is called for.

There are also types of experiment, especially with potted plants, where it is possible to gain statistical control of plant variability by allotting all treatments at random within size groups based on leaf area prior to application of treatments.

The technique described by McIntyre and Williams (1949) was designed for cases where the treatments are operative from germination, and where there are physical limits to the numbers of plants which can be handled per harvest class. They were seeking, for a field-grown tomato crop, levels of precision which are normally attainable only in pots in a controlled environment. The procedure is illustrated in Fig. A.10 for one harvest interval of the control series.

The difference between the mean weights of two harvests involves not only the sampling variation in the increments for individual plants between the two harvests, but also the deviation of the mean of the later harvested plants from the other set at the time of the first harvest. As the interval between harvests is shortened, this latter factor becomes progressively more important until, in the limit of zero increment, it is the sole source of error.

If it is possible to make objective measurements or ratings on both sets at the time of the first harvest, the importance of this factor can be reduced.

In our example it will be seen that we have information about 32 plants belonging to two randomly selected groups of 16 plants. One group was harvested on day 32 and their individual dry weights are plotted as a function of their leaf areas on day 31. The second group was harvested on day 39 and their weights are also plotted as a function of their leaf areas on day 31. The areas were estimated by a procedure which did not damage the plants.

Fig. A.10 presents the data both on absolute and logarithmic bases, and the weight-rating relations are seen to be adequately described by linear regressions. Intercepts with the common mean ratings supply estimates of dry weight at the beginning and end of the seven-day period of growth.

With reference to the logarithmic data, the mean rating on day 31 is 45.5

236

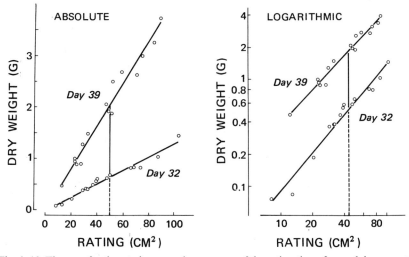

Fig. A.10. The use of ratings to improve the accuracy of the estimation of growth increments. (After McIntyre and Williams, 1949.)

cm² and the corresponding estimated mean log dry weights (g) using ratings are $\bar{1}.653$ and 0.204 for days 32 and 39 respectively. The increment is 0.551 ± 0.023. The standard error of the increment takes into account the error in the mean rating. The unadjusted mean log weights are $\bar{1}.643$ and 0.213 with increment of 0.570 ± 0.110. The relative standard errors of the increments show the very considerable gain in precision from rating.

Bouma and Dowling (1966) made very effective use of the rating procedure in their careful studies of plant response to nutrient stress. They used growth in leaf area as their index of response, so their bivariate distributions related to a common variate. They were able to establish differential responses within three days of the commencement of treatment.

Where the labour involved precludes the use of large numbers of replicates for the ultimate measurements, the use of median plants is recommended. We have found this to be necessary for all our studies of volume growth. For the flax experiment, in which there were harvests every three or four days, we harvested 50 plants per occasion by progressive random thinning. These plants were set out in a visual array according to height, and 12 'median' plants selected by rejecting 28 short and 10 tall plants. The 12 'median' plants differed very little in height but could be further classified by numbers of leaves in excess of a predetermined length. The odd plant that *looked* different by reason of slight differences of leaf shape or presentation was also excluded from the two most representative plants wanted for individual leaf length and fresh weight measurements. The other 10 were all fixed in formalin–acetic alcohol and later reduced to four or the basis of primordium length close to the apex. These four apices were duly embedded, with the object of securing two sets of serial sections which were sufficiently complete and correctly cut for volume integration. This elaborate procedure was dictated to some extent

237

by the need to link volume and fresh weight data for the same array of leaf primordia and leaves. Figures 3.1.15, 3.5.8, 3.6.13, 3.7.7 and 6.5 are all based on median-plant selection techniques of this kind, though usually less rigorous as to detail. Fuller accounts for wheat and clover will be found in Williams and Williams (1968) and Williams and Bouma (1970). Incidentally, the use of progressive thinning for growth studies not requiring the recovery of root systems is strongly recommended. It is economical of controlled-environment space, and need not introduce problems of self-shading if due account is taken of plant habit and pot size. It should not be necessary to stress that most studies of growth *per se* require controlled-environment facilities.

Anyone who has worked with vegetative apices will have been impressed by the near perfection of the pattern *within* an apex, so much so that they will automatically have seen in the succession of primordia a picture of the growth of a typical primordium. This impression is well founded, for we know that successive primordia are initiated at remarkably constant intervals of time and, during rapid early growth, they tend to establish exponential sequences with respect to length, volume, and of weight also (Fig. 5.4). These facts have been used to establish early-stage growth curves for flax (Fig. 4.1.17 and text), and for *Brassica* and *Lupinus* (Figs. 4.3.5 and 4.4.3). In each case a time scale had to be established independently, and caution was urged where final leaf size was known to increase with leaf number. The procedure is a powerful one for the recognition of special events such as the peak in growth activity during leaf initiation.

Finally we come to the concept of age equivalence. This was first developed in relation to the study of the dynamics of leaf growth in wheat (Chapter 5). During the period of rapid extension growth, leaf primordia are more variable in length than at any other time, and this poses a problem if precision is to be achieved in the quantitative description of changes in weight or in specific leaf constituents. Thus on day 16 (Fig. 5.7), the length of the fourth leaf ranged from 7 to 41 mm, and information would be lost if such variable material were pooled for analysis. The length–time relation of Fig. 5.7 was therefore used to determine the equivalent ages of primordia or leaves of given lengths. Thus the 7 mm primordium would have fallen within the range 6.77–7.47 mm and have had an age equivalent of 14.9 days. The 41 mm primordium would have been within the range 38.1–41.4 mm and have had an age equivalent of 16.8 days. This shows that the age equivalents of fourth leaves harvested on day 16 had a range of almost two days. Incidentally, the height ranges appropriate to every tenth of a day were read off from a large scale plot of the length–time relation on semi-log paper.

The way in which the procedure can be used to obtain daily estimates of dry weight is illustrated in Fig. A.11, where individual and group values are plotted against age equivalents. The estimates shown in the right-hand portion of the figure are mid-values from linear regressions for successive, overlapping two-day arrays of values. They are equivalent to running means in which the data are used twice.

Clearly, age equivalence is a powerful procedure for securing precise information about chemical and weight change in rapidly growing organs which

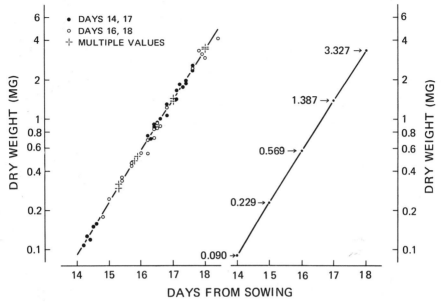

Fig. A.11. Diagram illustrating the use of age-equivalence for the determination of daily values of dry weight. For full explanation see text. (After Williams and Rijven, 1965.)

present sampling problems of the kind indicated. The reference variable – usually length – must itself be established with precision, for errors there will remain as hidden systematic errors in the dependent variable. Then, too, the possibility that primordia of the same length might differ in weight or composition if they attained that length at different times should be checked. Such effects would reveal themselves as systematic departures from general trend. They have not been detected in our work with wheat or clover. A great deal of the data of Chapter 4 is based on the use of age equivalence procedures.

The growth curves for *Ficus* (Fig. 4.8.11) would not have been possible without age equivalence, and Figs. 6.11 and 6.12 use a combination of median-plant and age-equivalence procedures to achieve their results. However, the most elaborate example in this book of the combination of procedures for the precise quantitative description of growth is that of Fig. 4.1.16. The details are given in the associated text.

A.5 DATA PROCESSING AND PRESENTATION

A special feature of many sets of data presented above is that they describe the growth of a series of like members within biological systems. This has prompted the use of a number of devices designed to bring out the fact that they *are* systems rather than collections of individuals. Figs. 2.5 and 4.1.14

illustrate the distinction quite well, for they present the same data for length growth of flax leaves. In the former, the representative leaves are independently fitted with Richards-type curves and this procedure is thought not to describe the system as well as the more subjective approach of Fig. 4.1.14. A somewhat hybrid procedure is used in Fig. 6.12, where the family of curves for stamen growth is based on independent quadratic regressions. For the carpels, the fuller knowledge of Fig. 6.11 is used to justify the somewhat subjective linking of a linear phase with a later curved segment (also quadratic) for florets 1, 2 and 3. The end result is a quantitative description which serves the purpose, though freehand curves would have served almost as well.

The sets of curves in Fig. 4.7.7 rest heavily on previous experience, and on a bare minimum of experimental values. This economy of effort was achieved by the simple device of harvesting on the days of successive leaf emergence, irrespective of treatment effects. In spite of the considerable effects of treatment on leaf number and rates of growth for comparable stages, all the systems are clearly built on a standard pattern of leaf growth. Any attempt at objective curve fitting would achieve less than the freehand curves which are in fact used.

The more orthodox treatment of Fig. 4.5.7 in which the fresh weight data for successive clover leaves are linked by straight lines, would gain little from a freehand treatment. However, the choice between the use of freehand curves and straight-line links will always remain, to a degree, a matter of personal preference. In this regard Figs. 4.1.11 and 4.1.12 provide a relevant contrast. Although there were 71 values in all for the radial distances of the former, these did not yield a satisfying picture when given straight-line links. There were usually three points appropriate to the selected primordia, but sometimes only two. With few exceptions, the sets of three were consistent with a slight negative curvature, and there was no obvious increase or decrease of curvature with primordium number. A very simple device was used to estimate the mean curvature – by superimposing an array of curves drawn on tracing film, and judging visually which of these gave the best fit. When establishing the family of curves, trends in neighbouring curves were taken into account. This permitted due allowance for the aberrant values for days 15 and 18. Many will query this degree of subjective treatment of the data, but I believe the implicit stress on the properties of the system is justified by the outcome.

By contrast, Fig. 4.1.12, with its straight-line linking of values, manages to convey the properties of the system without the need for special interpretive devices. This is doubtless because the rate of internode elongation becomes remarkably constant in the sub-apical region.

Several sets of data (Figs. 4.5.6, 4.5.8 and 4.7.9) lend themselves to the fitting of families of linear regressions. For example, the procedure for the length data of Fig. 4.5.6 is as follows, it being noted that there is little evidence of curvature in the log-length (L) data after day 4 and up to a length of 12 mm. Linear regressions of $\log L$ on time, t, were fitted independently for leaves 1–15 and it was found that their slopes changed progressively and significantly with leaf number, as did the abscissae at an arbitrary length (3.16 mm was used). A compound expression of the form $\log L = A + Bt$, with polynomial expressions in leaf number substituted for A and B was then fitted to the entire data, giving a multiple correlation coefficient of 0.9985. A fuller numerical statement of this exercise is given by Williams and Bouma (1970, Table 2).

A similar set of linear regressions was derived for the volumes of Fig. 4.5.8. The range of values covered by the regressions was 10^{-3} to about 5 mm³, but a graphical procedure based on the same principles was developed to produce the family of curves for growth over the range 2×10^{-5} to 10^{-3} mm³. Here the level of precision is much less than later, and the curves are shown as dotted lines.

Fig. 4.7.9 is a good example of the graphical approach to the fitting of families of linear regressions. These regressions were first fitted visually for each leaf independently, using a line drawn on tracing film. The slopes (b values) were plotted against primordium number and corrected on the assumption that they fall on a continuous curve. The corrected values (as relative growth rates) appear in Fig. 4.7.10, but not the original b values. There was not much scope for correcting by abscissae at any arbitrary length, so we had to be content with minor adjustments to some of the shorter regressions. Those for leaves 1 and 2 for treatment 8L, for instance, take account of the fact that all values for day 9 are too low.

In the preparation of Fig. 4.1.15 we were particularly anxious to avoid subjective bias in developing the required family of curves for leaf growth. The raw data consisted of volume and fresh weight values for selected leaves (every eighth, except at first) for the sixteen sampling occasions. Within occasion the volumes were not independent, being all derived from the same two axes, likewise the fresh weights were based on a single group of plants. While the rigorous median-plant approach to sampling produced a very good set of data, there was still opportunity for systematic error of a localized character.

The basis of the procedure adopted was that the times of attainment of a given volume by successive leaf primordia can be expected to be a continuous and fairly smooth function of leaf number. By interpolation and a limited amount of extrapolation (for the early values) we set up 12 such sequences of times, and then applied a system of step-wise or progressive curve fitting to each. The outcome is the family of curves of Fig. 4.1.15, and the method is exemplified in Table A.2 for part of the sequence for 100 units of volume.

Reading from the original large-scale plot on semi-log paper, it was noted that primordium 24 attained a volume of 100 units on day 19.4, primordium 32 did so on day 21.8, and so on to primordium 208 which attained 100 units of volume on day 56.5. The rest of Table A.2 will not need to be described in detail, but consists, in the main, of two sets of quartic regressions, each regression being fitted to a sequence of 7 to 10 values. The values shown in italics are means of terminal estimates of the appropriate regressions, and the regression coefficients are set out below. Incidentally these were rounded off to some extent and attempts to evaluate within the main body of the table may find disagreement in the second decimal.

There are many ways in which one could carry out such a programme of

Table A.2. *Progressive or stepwise fitting of polynomials to a sequence of times of attainment of 100 units of volume (see Fig. 4.1.15)*

Primordium no., x	Actual (day)	Set 1 (day)	Set 2 (day)	Average (day)	Difference (day)
			Estimated time, y		
24	19.4	—	19.37	—	—
32	21.8	—	21.92	—	—
40	24.6	24.62	24.44	24.53	−.07
48	26.6	26.56	26.63	26.59	−.01
56	28.3	28.29	28.40	28.34	.04
64	30.0	29.96	29.91	29.93	−.07
72	31.5	31.65	*31.49*	31.57	.07
80	33.4	33.38	33.51	33.44	.04
88	35.2	35.10	35.20	35.15	−.05
96	36.8	36.72	36.64	36.68	−.12
104	37.9	38.07	37.92	37.99	.09
112	39.0	*38.97*	39.14	39.05	.05
120	40.4	40.40	40.39	40.39	−.01
128	41.8	41.80	41.73	41.76	−.04
136	43.2	43.25	43.22	43.23	.03
144	44.9	44.81	*44.92*	44.88	−.02
152	46.5	46.48	46.36	46.42	−.08
160	48.1	48.27	48.28	48.27	.17
168	50.3	50.17	50.27	50.22	−.08
176	52.1	*52.19*	52.05	52.12	.02
184	53.5	53.34	53.48	53.41	−.09
192	54.5	54.42	54.57	54.49	−.01
200	55.5	55.50	55.47	55.48	−.02
208	56.5	56.58	56.50	56.54	.04

The Equations

Primordium sequence	a	b	c	d	e
		Set 1			
40–112	*[illegible]*	*[illegible]*	*[illegible]*	*[illegible]*	*[illegible]*
112–176	−59.530	2.4803	−.024733	.000114742	−.0000001921
176–208	28.500	0.1350			
		Set 2			
24–72	18.410	−0.4110	.028232	−.000450797	.0000024044
72–144	−28.219	1.6792	−.016668	.000075651	−.0000001138
144–208	1475.310	−33.3055	.285756	−.001072398	.0000014939

where $y = a + bx + cx^2 + dx^3 + ex^4$.

y is time in days, x is primordium number.

smoothing, and a careful plotting of the data to be smoothed will help in planning same. In this instance the average times provided a sound basis for calculating the relative growth rates of Fig. 4.1.18.

A.6 CELL COUNTING

This was undertaken only in relation to the two leaf studies of Chapter 5. For clover it was noted that there was only one nucleolus per nucleus, so for primordia between 0.1 and 3.0 mm in length, nucleoli were counted under oil immersion in 10 μm sections stained with haematoxylin and counter-stained with Orange G. In our material the nucleoli were black against a shadowy outline of the nucleus, and were seldom difficult to identify. From cumulative areas of the serial sections of a primordium, 32 sites were selected, and counting was done under an eyepiece micrometer grid. For small, relatively undifferentiated primordia, sampling areas were compact rectangles (12 grid squares), but for older primordia transect areas (up to 20×2 grid squares) were selected as objectively as possible to represent all tissues present. Between 300 and 600 nucleoli were counted per primordium, and the sampling ratio decreased from about 1 in 10 to 1 in 1000 with increase in leaf size.

If whole nuclei are to be counted, it is necessary to correct for their mean diameter or length. This should be estimated from longitudinal sections if the counts are made in transverse section. If the mean length be 7 μm and the counting is done in 10 μm sections, it follows that there will be 10 whole nuclei per section for every 17 counts based on recognizable fragments of nuclei.

A method appropriate to larger primordia is that of Rijven and Wardlaw (1966), in which the sample is stained in bulk with Feulgen's reagent, macerated by fungal cellulase, and the nuclei counted after deposition on a millipore filter. Counting is easier for wheat than for clover, for clover nuclei stain poorly. This difference is doubtless due to the fact that the DNA content of the clover nucleus is only one twenty-fifth of that of wheat.

A.7 CONVERSION TABLE

DOUBLING TIMES*

R	*Doubling time (days) as a function of R (day⁻¹)*									
	0	1	2	3	4	5	6	7	8	9
0.0	—	69.315	34.657	23.105	17.329	13.863	11.552	9.902	8.664	7.702
0.1	6.931	6.301	5.776	5.332	4.951	4.621	4.332	4.077	3.851	3.648
0.2	3.466	3.301	3.151	3.014	2.888	2.773	2.666	2.567	2.476	2.390
0.3	2.310	2.236	2.166	2.100	2.039	1.980	1.925	1.873	1.824	1.777
0.4	1.733	1.691	1.650	1.612	1.575	1.540	1.507	1.475	1.444	1.415
0.5	1.386	1.359	1.333	1.308	1.284	1.260	1.238	1.216	1.195	1.175
0.6	1.155	1.136	1.118	1.100	1.083	1.066	1.050	1.035	1.019	1.005
0.7	0.990	0.976	0.963	0.950	0.937	0.924	0.912	0.900	0.889	0.877
0.8	0.866	0.856	0.845	0.835	0.825	0.815	0.806	0.797	0.788	0.779
0.9	0.770	0.762	0.753	0.745	0.737	0.730	0.722	0.715	0.707	0.700

R	*Doubling time (hours) as a function of R (day⁻¹)*									
	0	1	2	3	4	5	6	7	8	9
1.0	16.64	15.12	13.87	12.80	11.88	11.10	10.40	9.79	9.24	8.76
2.0	8.32	7.92	7.57	7.23	6.93	6.65	6.40	6.16	5.94	5.74
3.0	5.55	5.37	5.20	5.04	4.89	4.75	4.62	4.50	43.8	4.27
4.0	4.16	4.06	3.96	3.87	3.78	3.70	3.62	3.54	3.47	3.40
5.0	3.33	3.26	3.20	3.14	3.08	3.03	2.97	2.92	2.87	2.82
6.0	2.77	2.73	2.68	2.64	2.60	2.56	2.52	2.48	2.45	2.41
7.0	2.38	2.34	2.31	2.28	2.25	2.22	2.19	2.16	2.13	2.11
8.0	2.08	2.05	2.03	2.00	1.98	1.96	1.93	1.91	1.89	1.87
9.0	1.85	1.83	1.81	1.79	1.77	1.75	1.73	1.72	1.70	1.68
10.0	1.66	1.65	1.63	1.62	1.60	1.58	1.57	1.56	1.54	1.53

R	*Doubling time (minutes) as a function of R (day⁻¹)*									
	0	1	2	3	4	5	6	7	8	9
10.0	99.81	90.74	83.18	76.78	71.30	66.54	62.38	58.71	55.45	52.53
20.0	49.91	47.53	45.37	43.40	41.59	39.93	38.39	36.97	35.65	34.42
30.0	33.27	32.20	31.19	30.25	29.36	28.52	27.73	26.98	26.27	25.59
40.0	24.93	24.35	23.77	23.21	22.69	22.18	21.70	21.24	20.80	20.37
50.0	19.96	19.57	19.20	18.83	18.48	18.15	17.82	17.51	17.21	16.92
60.0	16.64	16.36	16.10	15.84	15.60	15.36	15.12	14.90	14.68	14.47
70.0	14.26	14.06	13.86	13.67	13.49	13.31	13.13	12.96	12.80	12.64
80.0	12.48	12.32	12.17	12.03	11.88	11.74	11.61	11.47	11.34	11.22
90.0	11.09	10.97	10.85	10.73	10.62	10.51	10.40	10.29	10.19	10.08
100.0	9.98	9.88	9.79	9.69	9.60	9.51	9.42	9.33	9.24	9.16

* See pp. 13 and 223.

References

ALLSOPP, A. (1967). Heteroblastic development in vascular plants. *Advances in Morphogenesis* **6**, 127–71.

ARBER, A. (1934). *The Gramineae. A Study of Cereal, Bamboo and Grass,* Cambridge University Press.

AVERY, G. S. (1933). Structure and development of the tobacco leaf. *Am. J. Bot.* **20**, 565–92.

BALLARD, L. A. T. and PETRIE, A. H. K. (1936). Physiological ontogeny in plants and its relation to nutrition. 1. The effect of nitrogen supply on the growth of the plant and its parts. *Aust. J. exp. Biol. med. Sci.* **14**, 135–63.

BARNARD, C. (1955). Histogenesis of the inflorescence and flower of *Triticum aestivum* L. *Aust. J. Bot.* **3**, 1–20.

BARNARD, C. (1957). Floral histogenesis in the monocotyledons. 1. The Gramineae. *Aust. J. Bot.* **5**, 1–20.

BARNARD, C. (1964). Form and Structure. In *Grasses and Grasslands,* pp. 47–72, ed. C. Barnard, Macmillan: London.

BLACKMAN, V. H. (1919). The compound interest law and plant growth. *Ann. Bot.* **33**, 353–60.

BLACKMAN, V. H. (1920). The significance of the efficiency index of plant growth. *New Phytol.* **19**, 97–100.

BONNETT, O. T. (1936). The development of the wheat spike. *J. agric. Res.* **53**, 445–51.

BOUMA, D. and DOWLING, E. J. (1966). The physiological assessment of the nutrient status of plants. 2. The effect of the nutrient status of the plant with respect to phosphorus, sulphur, potassium, calcium or boron on the pattern of leaf area response following the transfer to different solutions. *Aust. J. agric. Res.* **17**, 633–46.

BROCK, T. D. (1967). Life at high temperatures. *Science* **158**, 1012–19.

BROUÉ, P., WILLIAMS, C. N., NEAL-SMITH, C. A. and ALBRECHT, L. (1967). Temperature and daylength response of some cocksfoot populations. *Aust. J. agric. Res.* **18**, 1–13.

CAUSTON, D. R. (1969). A computer program for fitting the Richards function. *Biometrics* **25**, 401–9.

CHURCH, A. H. (1904). *On the relation of phyllotaxis to mechanical laws,* Williams and Norgate: Oxford.

CLOWES, F. A. L. (1961). *Apical Meristems,* Blackwell Scientific Publications: Oxford.

CUTTER, E. G. (1965). Recent experimental studies of the shoot apex and shoot morphogenesis. *Bot. Rev.* **31**, 7–13.

DENNE, M. P. (1966a). Morphological changes in the shoot apex of *Tri-*

245

References

folium repens. L. 1. Changes in the vegetative apex during the plastochron. *N.Z. Jl Bot.* **4**, 300–14.

DENNE, M. P. (1966*b*). Leaf development in *Trifolium repens. Bot. Gaz.* **127**, 202–10.

DORMER, K. J. (1972). *Shoot Organization in Vascular Plants,* Syracuse University Press.

ELLIS, E. L. and DELBRUCK, M. (1939). The growth of bacteriophage. *J. gen. Physiol.* **22**, 365–83.

ERICKSON, R. O. (1973). Tubular packing of spheres in biological fine structure. *Science* **181**, 705–16.

ERICKSON, R. O. and MICHELINI, F. J. (1957). The plastochron index. *Am. J. Bot.* **44**, 297–305.

ESAU, K. (1953). *Plant Anatomy,* Wiley: New York.

EVANS, G. C. (1972). *The quantitative analysis of plant growth,* Blackwell Scientific Publications: Oxford.

FRIEND, D. J. C., HELSON, V. A. and FISHER, J. E. (1962). The rate of dry weight accumulation in Marquis wheat as affected by temperature and light intensity. *Can. J. Bot.* **40**, 939–55.

FUJITA, T. (1938). Statistiche Untersuchung über die Zahl der konjugierten Parastichen bei den schraubigen Organstellungen. *Bot. Mag., Tokyo* **52**, 425–33.

GATES, C. T., WILLIAMS, W. T. and COURT, R. D. (1971). Effects of droughting and chilling on maturation and chemical composition of Townsville stylo (*Stylosanthes humilis*). *Aust. J. agric. Res.* **22**, 369–81.

GATES, C. T., HAYDOCK, K. P. and WILLIAMS, W. T. (1973). A study of the interaction of cold stress, age, and phosphorus nutrition on the development of *Lotononis bainesii* Baker. *Aust. J. biol. Sci.* **26**, 87–103.

GILL, W. R. and MILLER, R. D. (1956). A method for study of the influence of mechanical impedance and aeration on the growth of seedling roots. *Soil Sci. Soc. Amer. Proc.* **20**, 154–7.

GREACEN, E. L. and OH, J. S. (1972). Physics of root growth. *Nature, Lond.* **235**, 24–5.

GREGORY, F. G. (1956). General aspects of leaf growth. In *The Growth of Leaves,* pp. 3–31, ed. F. L. Milthorpe, Butterworths Scientific Publications: London.

GREGORY, R. A. and ROMBERGER, J. A. (1972). The shoot apical ontogeny of the *Picea abies* seedling. 1. Anatomy, apical dome diameter, and plastochron duration. *Amer. J. Bot.* **59**, 587–97.

HACKETT, C. (1973). An exploration of the carbon economy of the tobacco plant. 1. Inferences from a simulation. *Aust. J. biol. Sci.* **26**, 1057–71.

HAGERUP, O. (1930). Vergleichende morphologische und systematische Studien über die Ranken und andere vegetative Organe der Cucurbitaceen und Passifloraceen. *Dansk bot. Ark.* **6**, 1–103.

HANNAM, R. V. (1968). Leaf growth and development in the young tobacco plant. *Aust. J. biol. Sci.* **21**, 855–70.

HIS, W. (1888). On the principle of animal morphology. *Proc. R. Soc. Edinb.* **15**, 287–98.

HOPKINSON, J. M. and HANNAM, R. V. (1969). Flowering in tobacco: The course of floral induction under controlled conditions and in the field. *Aust. J. agric. Res.* **20**, 279–90.

HOWE, K. J. and STEWARD, F. C. (1962). Studies in *Mentha piperita* L. 2. Anatomy and development of *Mentha piperita* L. New York State College of Agriculture, *Memoir* **379**, 11–40.

HUSSEY, G. (1971). Cell division and expansion and resultant tissue tensions in the shoot apex during the formation of a leaf primordium in tomato. *J. exp. Bot.* **22**, 702–14.

HUSSEY, G. (1972). The mode of origin of a leaf primordium in the shoot apex of the pea (*Pisum sativum*). *J. exp. Bot.* **23**, 675–82.

HUXLEY, J. (1954). The evolutionary process. In *Evolution as a process*, pp. 1–23, eds. J. Huxley, A. C. Hardy and E. B. Ford, Allen and Unwin: London.

JAMES, D. B. and HUTTO, J. M. (1972). Effects of tiller separation and root pruning on the growth of *Lolium perenne* L. *Ann. Bot.* **36**, 485–95.

KIDD, F., WEST, C. and BRIGGS, G. E. (1920). What is the significance of the efficiency index of plant growth. *New Phytol.* **19**, 88–96.

LEIGH, E. G. (1972). The golden section and spiral leaf-arrangement. *Connecticut Acad. Arts Sci. Trans.* **44**, 163–76.

LYNDON, R. F. (1968). Changes in volume and cell number in the different regions of the shoot apex of *Pisum* during a single plastochron. *Ann. Bot.* **32**, 371–90.

LYNDON, R. F. (1970). Rates of cell division in the shoot apical meristem of *Pisum. Ann. Bot.* **34**, 1–17.

MACDOWALL, F. D. H. (1972). Growth kinetics of Marquis wheat. 1. Light dependance. *Can. J. Bot.* **50**, 89–99. 2. Carbon dioxide dependance. *Ibid.* **50**, 883–9. 3. Nitrogen dependance. *Ibid.* **50**, 1749–61.

MACDOWALL, F. D. H. (1973). 4. Temperature dependance. *Ibid.* **51**, 729–36.

MCINTYRE, G. A., GRASSIA, A. and WARD, M. W. (1971). *Miscellaneous programs in the fields of analysis, linear and nonlinear regression, biological assay, association and frequency distributions. C.S.I.R.O. (Australia) Division of Mathematical Statistics. Tech. Report* No. 5, 21–3.

MCINTYRE, G. A. and WILLIAMS, R. F. (1949). Improving the accuracy of growth indices by the use of ratings. *Aust. J. scient. Res. B.* **2**, 319–45.

MAKSYMOWYCH, R. (1973). *Analysis of leaf development*, Cambridge University Press.

MILTHORPE, F. L., ed. (1956). *The Growth of Leaves*, Butterworths Scientific Publications: London.

NĚMEC, B. (1903). Ueber den Einfluss der mechanischen Factoren auf die Blattstellung. *Bull. Int. Acad. Sci. Boheme*, no. 8, 65–79.

OTSUKI, Y., SHIMOMURA, T. and TAKEBE, I. (1972). Tobacco mosaic virus multiplication and expression of the N gene in necrotic responding tobacco varieties. *Virology* **50**, 45–50.

PATTEE, H. H. (1970). The problem of biological heriarchy. In *Towards a theoretical biology*, 3. *Drafts*, pp. 117–36, ed. C. H. Waddington, Edinburgh University Press.

247

References

PATTEE, H. H., ed. (1973). *Hierarchy Theory, The Challenge of Complex Systems*, George Braziller: New York.

PETRIE, A. H. K. (1937). Physiological ontogeny in plants and its relation to nutrition. 3. The effects of nitrogen supply on the drifting composition of the leaves. *Aust. J. exp. Biol. med. Sci.* **15**, 385–404.

PLANTEFOL, L. (1948). *La Theorie des Helices Foliaires Multiples*, Masson et Cie: Paris.

RICHARDS, F. J. (1948). The geometry of phyllotaxis and its origin. *Symp. Soc. exp. Biol.* **2**, 217–45.

RICHARDS, F. J. (1951). Phyllotaxis: its quantitative expression and relation to growth in the apex. *Phil. Trans. R. Soc. B.* **235**, 509–64.

RICHARDS, F. J. (1956). Spatial and temporal correlations involved in leaf pattern production at the apex. In *The Growth of Leaves*, pp. 66–76, ed. F. L. Milthorpe, Butterworths Scientific Publications: London.

RICHARDS, F. J. (1959). A flexible growth function for empirical use. *J. exp. Bot.* **10**, 290–300.

RICHARDS, F. J. (1969). The quantitative analysis of growth. In *Plant Physiology, A Treatise*, vol. 5A, pp. 3–76, ed. F. C. Steward, Academic Press: New York and London.

RICHARDS, F. J. and SCHWABE, W. W. (1969). Phyllotaxis: A problem of growth and form. In *Plant Physiology, A Treatise*, vol. 5A, pp. 79–116, ed. F. C. Steward, Academic Press: New York and London.

RIJVEN, A. H. G. C. (1968). Determination of uni- and trifoliolate leaf form in fenugreek. *Aust. J. biol. Sci.* **21**, 155–6.

RIJVEN, A. H. G. C. and WARDLAW, I. F. (1966). A method for the determination of cell number in plant tissues. *Exptl Cell Res.* **41**, 324–8.

ROSEN, R. (1967). *Optimality principles in biology*, Butterworths: London.

SCHWABE, W. W. (1971). Chemical modifications of phyllotaxis and its implications. In *Control Mechanisms of Growth and Differentiation, Symp. Soc. exp. Biol.* **25**, 301–22.

SHARMAN, B. C. (1942). Developmental anatomy of the shoot of *Zea mays* L. *Ann. Bot. N.S.* **6**, 245–82.

SHARMAN, B. C. (1943). Tannic acid and iron alum with safranin and orange G in studies of the shoot apex. *Stain Technol.* **18**, 105–11.

SHARMAN, B. C. (1945). Leaf and bud initiation in the Gramineae. *Bot. Gaz.* **106**, 269–89.

SNOW, M. (1951). Experiments on spirodistichous shoot apices. *Phil. Trans R. Soc. B.* **235**, 131–62.

SNOW, M. and SNOW, R. (1931). Experiments on phyllotaxis. 1. The effect of isolating a primordium. *Phil. Trans. R. Soc. B.* **221**, 1–43.

SNOW, M. and SNOW, R. (1935). Experiments on phyllotaxis. 3. Diagonal splits through decussate apices. *Phil. Trans. R. Soc. B.* **225**, 63–94.

SNOW, M. and SNOW, R. (1947). On the determination of leaves. *New Phytol.* **46**, 5–19.

SNOW, R. (1952). On the shoot apex and phyllotaxis of *Costus*. *New Phytol.* **51**, 359–63.

248

SNOW, R. (1955). Problems of phyllotaxis and leaf determination. *Endeavour* **14**, 190–99.

SNOW, R. (1965). The causes of the bud eccentricity and the large divergence angles between leaves in Cucurbitaceae. *Phil. Trans. R. Soc. B.* **250**, 53–77.

STEWARD, F. C. (1968). *Growth and Organization in Plants*, Addison-Wesley Publishing Co.: Reading, Massachusetts.

SUNDERLAND, N., HEYES, J. K. and BROWN, R. (1956). Growth and metabolism in the shoot apex of *Lupinus albus*. In *The Growth of Leaves*, pp. 77–90, ed. F. L. Milthorpe, Butterworths Scientific Publications: London.

THOMPSON, D'ARCY W. (1942). *On Growth and Form* (2nd edition), p. 920, Cambridge University Press.

TRINCI, A. P. J. (1969). A kinetic study of the growth of *Aspergillus nidulans* and other fungi. *J. gen. Microbiol.* **57**, 11–24.

TURPIN, P. J. F. (1819). Mémoire sur l'inflorescence des Graminées et des Cypéracées. *Mém. du mus. d'hist. nat., Paris* **5**, 426–92.

VAN ITERSON, G. (1907). *Mathematische und mikroscopisch – anatomische Studien über Blattstellungen*, Jena.

VIDAVER, W. (1972). Effects of pressure on the metabolic processes of plants. In *The Effects of Pressure on Organisms, Symp. Soc. exp. Biol.* **26**, 159–74.

WARDLAW, C. W. (1949). Experimental and analytical studies of Pteridophytes. XIV. Leaf formation and phyllotaxis in *Dryopteris aristata* Druce. *Ann. Bot.* **13**, 163–98.

WARDLAW, C. W. (1952). *Phylogeny and Morphogenesis*, Macmillan: London.

WARDLAW, C. W. (1965). The organization of the shoot apex. In *Encyclopedia of Plant Physiology* XV/1, pp. 966–1076, ed. W. Ruhland, Springer-Verlag: Berlin.

WAREING, P. F. and PHILLIPS, I. D. J. (1970). *The Control of Growth and Differentiation in Plants*, Pergamon Press: Oxford.

WEISS, P. A. (1969). The living system: determinism stratified. In *The Alpbach Symposium, 1968. Beyond Reductionism – New Perspectives in the Life Sciences*, pp. 3–55, eds A. Koestler and J. R. Smythies, Hutchinson: London.

WEISSE, A. (1932). Zur Kenntnis der Blattstellungsverhaltnisse bei den Zingiberaceen. *Ber. dtsch. bot. Ges.* **50a**, 327–66.

WHALEY, W. G. (1961). Growth as a general process. In *Encyclopedia of Plant Physiology* XIV, pp. 71–112, ed. W. Ruhland, Springer-Verlag: Berlin.

WHITEHEAD, A. N. (1933). *Science and the modern world*, Cambridge University Press.

WILLIAMS, R. F. (1955). Redistribution of mineral elements during development. *Ann. Rev. Pl. Physiol.* **6**, 25–42.

WILLIAMS, R. F. (1960). The physiology of growth in the wheat plant. 1. Seedling growth and the pattern of growth at the shoot apex. *Aust. J. biol. Sci.* **13**, 401–28.

WILLIAMS, R. F. (1964). The quantitative description of growth. In *Grasses and Grasslands*, pp. 89–101, ed. C. Barnard, Macmillan: London.

WILLIAMS, R. F. (1966a). The physiology of growth in the wheat plant. 3.

References

Growth of the primary shoot and inflorescence. *Aust. J. biol. Sci.* **19**, 949–66.

WILLIAMS, R. F. (1966*b*). Development of the inflorescence in Gramineae. In *The Growth of Cereals and Grasses*, pp. 74–87, ed. F. L. Milthorpe, J. D. Ivins, Butterworths: London.

WILLIAMS, R. F. (1970). The genesis of form in flax and lupin as shown by scale drawings of the shoot apex. *Aust. J. Bot.* **18**, 167–73.

WILLIAMS, R. F. and BOUMA, D. (1970). The physiology of growth in sub-terranean clover. 1. Seedling growth and the pattern of growth at the shoot apex. *Aust. J. Bot.* **18**, 127–48.

WILLIAMS, R. F. and LANGER, R. H. M. (1975). The physiology of growth in the wheat plant. 6. The dynamics of tiller growth. *Aust. J. Bot.* **23** (in press).

WILLIAMS, R. F. and METCALF, R. A. (1975). Physical constraint and tiller growth. *New Phytol.* **74** (in press).

WILLIAMS, R. F. and RIJVEN, A. H. G. C. (1965). The physiology of growth in the wheat plant. 2. The dynamics of leaf growth. *Aust. J. biol. Sci.* **18**, 721–43.

WILLIAMS, R. F. and RIJVEN, A. H. G. C. (1970). The physiology of growth in subterranean clover. 2. The dynamics of leaf growth. *Aust. J. Bot.* **18**, 149–66.

WILLIAMS, R. F., SHARMAN, B. C. and LANGER, R. H. M. (1975). The physiology of growth in the wheat plant. 5. Growth and form of the tiller bud. *Aust. J. Bot.* **23** (in press).

WILLIAMS, R. F. and WILLIAMS, C. N. (1968). Physiology of growth in the wheat plant. 4. Effects of daylength and light energy level. *Aust. J. biol. Sci.* **21**, 835–54.

WOODGER, J. H. (1929). *Biological Principles*, Kegan Paul: London. (Reissued in 1967, Routledge and Kegan Paul: London.)

WRIGHT, C. (1873). The uses and origin of the arrangements of leaves in plants. *Mem. Amer. Acad.* **9**, 379–418.

WRIGHT, S. T. C. (1961). Growth and cellular differentiation in the wheat coleoptile. 1. Estimation of cell number, cell volume and certain nitrogenous constituents. *J. exp. Bot.* **12**, 303–18.

Author and subject indexes

Author index

(*Italic type refers to text figures*)

Albrecht, L., 25
Allsopp, A., 118
Arber, A., 185, 213
Avery, G. S., 81

Ballard, L. A. T., 1
Barnard, C., *21*, *22*, *23*, *24*, 131, 183, 192, 193, 227
Blackman, V. H., 9
Bonnett, O. T., 183
Bouma, D., 5, 103, *103*, *105*, 108, 110, *110*, *111*, *112*, 113, *113*, *114*, *115*, 144, 171, 237, 238, 241
Briggs, G. E., 9
Brock, T. D., 13
Broué, P., 25, *26*
Brown, R., 99

Causton, D. R., 17
Church, A. H., 27, 35, 42, 51
Clowes, F. A. L., 7, 215
Court, R. D., 25
Cutter, E. G., 220

Delbruck, M., 13
Denne, M. P., 101, 107, *108*, 109, 218
Dormer, K. J., 7, 9, 17, 27, 44

Ellis, E. L., 13
Erickson, R. O., 43, 129
Esau, K., 58
Evans, G. C., 7

Fisher, J. E., 25
Friend, D. J. C., 25
Fujita, T., 31

Gates, C. T., 25
Gill, W. R., 212
Grassia, A., 17
Greacen, E. L., 212
Gregory, F. G., 192
Gregory, R. A., 68

Hackett, C., 3
Hagerup, O., 107
Hannam, R. V., 81, 88, *88*, 89, 90, 144, 230

Haydock, K. P., 25
Helson, V. A., 25
Heyes, J. K., 99
His, W., 4, 227
Hopkinson, J. M., 88, *88*
Howe, K. J., *127*, 128
Hussey, G., 217, 218, 219, 220
Hutto, J. M., 206
Huxley, J., 5

James, D. B., 206

Kidd, F., 9

Langer, R. H. M., 135, 199, 201, 213
Leigh, E. G., 44
Lyndon, R. F., 215, 218, 219, *219*

Macdowall, F. D. H., 10, 24, 25
McIntyre, G. A., 17, 236, *237*
Maksymowych, R., 8
Metcalf, R. A., 199, 204
Michelini, F. J., 129
Miller, R. D., 212
Milthorpe, F. L., 7

Neal-Smith, C. A., 25
Nemec, B., 213

Oh, J. S., 212
Otsuki, Y., 13

Pattee, H. H., 6
Petrie, A. H. K., 1
Phillips, I. D. J., 8
Plantefol, L., 220

Richards, F. J., 7, 16, 17, *18*, 27, 32, *32*, 33, *33*, 35, 37, 39, *40*, 41, *41*, 42, 43, 52, 75, 110, 136, 207, 220, 230
Rijven, A. H. G. C., 5, 79, 131, 144, 163, *164*, *165*, *166*, *167*, 169, *169*, 170, *170*, 171, *173*, 174, *174*, *175*, *176*, *177*, *178*, *179*, 180, 214, *239*, 243
Romberger, J. A., 68
Rosen, R., 7, 209

251

Schwabe, W. W., 27, 221
Sharman, B. C., 131, 135, 176, 192, 199, 201, 213, 218, 224
Shimomura, T., 13
Snow, M., 27, 106, 216, 221
Snow, R., 27, 64, 107, 213, 216, 220, 221
Steward, F. C., 7, 9, 17, *127*, 128
Sunderland, N., 99

Takebe, I., 13
Thompson, D'Arcy W., 27, 31, 43, 44
Trinci, A. P. J., 13
Turpin, P. J. F., 185, 213

Van Iterson, G., 42, 43
Vidaver, W., 6

Ward, M. W., 17
Wardlaw, C. W., 27, 220
Wardlaw, I. F., 243
Wareing, P. F., 8

Weiss, P. A., 3
Weisse, A., 213
West, C., 9
Whaley, W. G., 8, 13
Whitehead, A. N., 4
Williams, C. N., 5, 25, 131, 136, *138*, 139, *139*, 140, 142, 144, 238
Williams, R. F., 2, 5, 13, 17, 79, 97, *98*, 103, *103*, *105*, 108, 110, *110*, *111*, *112*, 113, *113*, *114*, *115*, 131, *133*, *134*, 135, 136, *138*, 139, *139*, 140, 142, 144, 163, *164*, *165*, *166*, *167*, 169, *169*, 170, *170*, 171, *173*, 174, *174*, *175*, *176*, *177*, *178*, *179*, 180, 183, *184*, *187*, *188*, *189*, *190*, *191*, 193, *193*, *194*, *195*, *196*, *197*, 198, 199, 201, 204, 210, 213, 223, *224*, *225*, *226*, *228*, 236, *237*, 238, *239*, 241
Williams, W. T., 25
Woodger, J. H., 4, 220
Wright, C., 43
Wright, S. T. C., 175

Subject index

abstraction, intolerant use of, 4
Acacia mucronata, 34, 158, *159*, 160, *160*
age equivalence, 238–9, *239*
Agropyron, 218
algae, growth rate, 13
Anabena cylindrica, growth rate, 13
apex–primordium area ratio, 39, *41*
  in: cauliflower, 94; flax, 66; tobacco, 97
apical cone, 28, *29*, 35, 137
apical dome, 34–5, *34*
  in: cauliflower, *92*, 218; clover, *103*, 107–8; eucalyptus, 129; flax, *57*–9, 58–9; serradella, *157*, 158; tobacco, 83–4, *83*–5, 86; wheat, *132*–3, 137, *145*
Araucaria excelsa, 33
area ratio, *see* apex–primordium area ratio
Aspergillus nidulans, growth rate, 13
autocatalysis, curve of, 18, 20, *21*
Avena, 183
axial growth in: flax, 70–3, 208; wheat, 139–40, 210
axillary bud in: clover, 113, *114*, 115–16, *115*; eucalyptus, *120*, 121, *124*; wheat, 133, 135, *135*; *see also under* tillers

Bacillus stearothermophilus, growth rate, 13
bacteria, growth rate, 13
bacteriophage, growth rate, 13
Brassica oleracea, 29, *30*, 35–6, *36*, *41*, 91–7, 210, 211, 214–18, *215*, 232–3, *233*, 234, *234*

carpel, 185, *188*, *190*, *191*, 192–3, 195, *195*, 196–7, *196*, 197
cauliflower, *see Brassica oleracea*
cell
  counting, 243
  mass flow, 218–19, *219*
  number per leaf, 165–6, *166*, 174
  volume, 170–1, 175, 177
cell-wall materials in: clover leaf, *167*, 169–72, *170*; wheat leaf, 172, 176–7, *176*
Chlorella, growth rate, 13
chrysanthemum, phyllotactic change, 221
clover, *see Trifolium*
cocksfoot, *see Dactylis glomerata*
corpus tissue
  growth centres in, 220
  growth rate of, 113, *114*, 115, *115*, 116
Costus cylindrica, 213

covariance, with ratings, 236–7, *237*
curve fitting, 16–24
    exercise in, 20–4
    families of curves, 240–3
    polynomials, 20, *21*
    progressive, 20, 241–2

Dactylis glomerata, 25, *26*
data processing, 239–42
developmental space, 132, 155, 211
Dianella, 30, *30*, 136, 157–8, *159*, 221
divergence angle, 35–6, *36*, 38
    in: clover, 103, 106; flax, 66, *67*; tobacco, 87
DNA-phosphorus
    in: clover leaf, *167*, 169–70, *170*, 179; wheat leaf, *176*, 178
    per cell, 166, 174
doubling time, 13
    conversion table, 244
    in: clover, *113*, *114*, *179*; flax, *79*; tobacco, *90*; wheat, 139, 141, *178*, *204*
dry weight change in: clover, 168; wheat, *174*

Ecballium elaterium, 107
efficiency index, 9; *see also* relative growth rate
Epilobium, surgical experiments on, 221
equivalent phyllotaxis index (E.P.I.), 39
Escherichia coli, growth rate, 13
Eucalyptus, 116–31, 210, 211, 230
    juvenile and mature leaves of, 118, *125*, *127*
    primordium deformation in, 119, *122*, *123*, 213
Eucalyptus bicostata, 30, *30*, 116, *117*, 118–19, *119*, *120*, 121, *122*, 123, *123*, *124*, *125*, *126*, *127*, *128*, 129–30, *130*
Eucalyptus grandis, 116, *117*, 118, *120*, 128, *131*
Euphorbia, 34
exponential growth, 22, *23*, 24, *24*, 25
    in: clover, 109, 111–12, *113*, *114*, *115*, 116, 208–9; eucalyptus, 130; fig, 155–6; tobacco, 210; wheat, 141–2, 144, 209
    not in tiller buds, 202
Fibonacci angle, 42–55
    in: fig, 217; flax, 66–7; tobacco, 86–7
    optimal leaf display, 43–4
Fibonacci series, 29
Ficus elastica, *41*, 146–56, 21Î, 213, 217
fig, *see Ficus elastica*
flax, 207–8
    methods based on, 231, *231*, 232, 233, *233*, 234–5, *235*, 238, 241–2
    see also Linum usitatissimum

floret, 183, *186–7*, *188*, *189*, *191*, 192, 195–7, *196–7*
form change
    of leaf primordia in: *Acacia*, 158, *159*, 160, *160*; cauliflower, 91, *92*; clover, 101–2, *102*, *103*, 163–4, *164*; *Dianella*, 157–8, *159*; eucalyptus, 119, *119*, *120*, 121, *122*, 123, *123*; fig, 149, 151–3, *152*, *153*, *154*; lupin, 97, *98*; serradella, 157, *158*; tobacco, 81–6, *82*, *83*, *84*; wheat, 131–2, *132*, 144–6, *145*
    of tiller buds, 133, 135, *135*, 199–201, *200*, *202*, 213–14
    see also genesis of form
fresh weight change in: clover, 110–11, *111*, *112*, 168; flax, 75, *76*, 77
fungi, growth rate, 13

generation time, *see* doubling time
genesis of form, 212–14; *see* form change
genetic spiral, 36, *36*, 91
geometrical modelling of phyllotaxis
    Fibonacci system: orthogonal 2:3, 46–8, *47*, *49*, *50*; orthogonal 3:5, 44–6, *45*, 48; orthogonal 5:8, 48; progression from low to high, 49, *50*, 51–2, *51*
    first accessory system: orthogonal 3:4, 53–4, *54*; orthgonal 4:7, 52–3, *53*
    bijugate system, orthogonal 4:6, 54–5, *55*
glumes, 183, *187*, *188*, *189*, *191*, 192, 194, *194*, 195
gnomon, to bare apex, 35
'golden' section, 125, *127*
Gompertz function, 18
growth
    quantitative description of, 9–26
    see also: axial growth; exponential growth; inflorescence growth; leaf growth; length growth; tillers, growth of; and volume growth
growth analysis, 10, 25, *26*
'growth coefficient', 10
growth curves
    analysis of, 10, *11*, 12
    composite in: clover, 110–11, *112*; eucalyptus, *130*, *131*; fig, *156*; flax, 75, *76*, 77; wheat, *138*
    Richards's function, 17–18, *18*, *19*, 20
growth hormones and phyllotaxis, 221

hairs, epidermal
    effects on growth: in lupin, 99; and in tobacco, 81, *85*, 211
    glandular, in *Acacia*, 160, *160*
    growth rate, in clover, *166*, 171–2, 181
Helianthus annuus, 29, *30*, 31, 41, *41*, 160, *161*, 162

Subject index

hierarchic order and biology, 3–4, 6
histogenesis, foliar and cauline, 183, 192, 197–8
'holism' in biology, 3

inflorescence growth in wheat, *138*, *139*, 183–98
internode growth in: flax, 70–3, *71*, *73*; wheat, 139–40

leaf growth, 163–82
   in: cauliflower, 94–7; clover, 108–16, 167–72; eucalyptus, 129–31; fig, 153–6; flax, 73–80; lupin, 98–100; tobacco, 89–91; wheat, 140–6, 172–7
lemma, 183, *188*, *189*, *191*, 192, 194, *194*, 195, 196, *196*, 210
length growth
   in: clover, 108–9, *110*, 164–5, *165*, 168; eucalyptus, *128*, 129–30; flax, 72, *74*, 75, 77–8, *77*; wheat, 172–4, *173*
Linum usitatissimum, 13, *15*, *16*, 28, 29, *29*, *30*, *41*, 56–80, 224–7, *225*, *226*, 229, *229*; *see also* flax
lodicule, 185, *191*, 192
logarithmic scales, use of, 10, *11*, 12
Lolium perenne, 206
longitudinal section
   diagram: for cauliflower, *215*; for clover, *108*, *114*; and for wheat, *142*, *143*
   photomicrograph: for eucalyptus, *124*; for flax, *61*; and for wheat, *133*
Lotononis bainesii, 25
Lufa cylindrica, 107
lupin, *see Lupinus*
Lupinus albus, 64
Lupinus angustifolius, 97–100, 210, 211

marginal meristems in: cauliflower, 91, *92*, 97, 211; clover, 101, *103*; fig, 155–6; flax, 59, 80; lupin, 99, 211; serradella, 157
median-plant techniques, 23/–8
Mentha piperita, *127*, 128–9, 232
mitotic index in clover apex, 108
monomolecular curve, 18
multivariate analysis, 25–6

Nicotiana tabacum, *41*, 81–91; *see also* tobacco
nucleic acids in: clover leaves, *167*, *169*, 169–71, *170*; wheat leaves, 174–7, *176*, *177*
nucleoli, size of, 216
optimality in development, 5, 7, 209, 212
Ornithopus compressus, 157, *158*; *see also* serradella
ovary (and ovule), 185, *191*, *196*

palea, 183, *191*, 192, 195–6, *195*, *196*, 210
parastichies, 28
   Church's system of, 28
   contact, 28–34
   contact in: cauliflower, 91, *96*; fig, 153, *155*; flax, *63*, 64–5; tobacco, 86–7, *87*
   intersection of, 34, *40*
pea, 218–20; *see also Pisum sativum*
percentage cover, 66
   in: cauliflower, 94; flax, 66
petiole in: clover, 101, 163–4, *164*; lupin, 97, *98*
phyllotactic system
   alternate, 38
   bijugate, 29, 38
   decussate, 30, *30*, 38, 42, 63, 116–18, *117*
   distichous, 29–30, *30*, 42, 136
   Fibonacci spiral, 29, *30*, 31, 36–8, 40, 160, *161*, 162; fully defined by plastochrone ratio, 70; progression through, 65
   first accessory, 29, 38
   multijugate, 29
   spiral, 28–35, 64, 86
   spirodistichous, 106, 213
   whorled, 38, 42
phyllotaxis, 7, 27–55
   definition, 28
   history, 27
   in: *Acacia*, 160; cauliflower, 91–4; clover, 103, 105–7; *Dianella*, 157; eucalyptus, 123, 125, 127–9; fig, 153, *155*; flax, 63–70; lupin, 97–8; sunflower, 160–2; tobacco, 86–9; wheat, 136–8
   independent of primordium shape, 32–5
   measurement, 230–5
   parameters, 35–42
   Richards's procedures, 27, 41–2
   transverse component, 35
phyllotaxis index (P.I.), 37–8, *41*
   of bijugate systems, 38
   of first accessory systems, 38
   in: cauliflower, 91, 94; clover, 106; *Dianella*, 136, 157; eucalyptus, 128; flax, 65–6; *Mentha*, 129; sunflower, 162, tobacco, 87; wheat, 137
phyllotaxis theory, 220–1
   field theory, 216–17, 220
   first available space, 216–17, 220
   mechanico-chemical field theory, 220–1
physical constraint
   as determinant in biology, 5
   by epidermal hairs, 81, *85*, 99, 211
   of floral parts, 185; of roots, 212; of tiller buds, 199–206
   and genesis of form, 207, 212–14, 216–17
   and growth rate, 207–12
   and hierarchical control, 6

254

physical constraint (*cont.*)
in: cauliflower, 91, 97; eucalyptus, 121, 130, 213; fig, 156; flax, 58, 60, 63, 75; wheat, 133–5, 144
Picea abies, 68
Pisum sativum, 215, 219; *see also* pea
plant response, study of, 25
plastochrone, 35, 67
in: cauliflower, 215; clover, 109; flax, 67–8, *68*; tobacco, 88
plastochrone index, 129
plastochrone ratio (*r*), 35–6
estimation of, 230, 232–5, *234*, *235*
in: cauliflower, 91, 94; clover, 106; *Dianella*, 136; eucalyptus, 127–8; flax, 65, 66; lupin, 97; *Mentha*, 128; sunflower, 160, 162; tobacco, 87; wheat, 137
primordia
packing of floral, 185
packing of foliar in: clover, *102*, 103, *104*, *105*, 113, *114*; eucalyptus, 123, *125*, *126*, 130; fig, 146, *148*, 149, *150*; flax, *57*, 58, *58*, 62, 75, 80, *80*; wheat, 132, 144–5, *145*
primordium attachment area, 28, *29*
progressive thinning for growth studies, 238
protein nitrogen in: clover leaf, *167*, 169–71, *169*, *170*; wheat leaf, 174–7, *174*, *175*
protophloem differentiation, timing of, 144

quantitative biology
climate of scientific opinion and, 3–4
integration and, 7
magnitude of task, 3
need for, 3–5
Plato and, 1
and size, 3

radial growth rate in: flax, 59, 69–70, *69*; wheat, *140*, 142–3; *see also* relative growth rate, radial
ratings, *see* covariance
reconstruction, three dimensional
for: *Acacia*, *159*; apical cone, *29*; cauliflower, *92*; clover, *102*, 103, *164*; *Dianella*, *159*; eucalyptus, *119*, *120*, *122*, *123*; fig, *152*, *153*, *154*; flax, *57*, *58*, *59*; lupin, *98*, *99*; serradella, *158*; tobacco, *82*, *83*, *84*; wheat, *132*, *135*, *145*, *190*, *200*, *202*
method, 223–7
redistribution of elements, 2
'reductionism' in biology, 3
relative growth rate, 10–14
axial: in flax, 70–3, *73*; and in wheat, 138–40, 143

compound interest law and, 9–10
conversion to doubling time, 13, 244
in: cauliflower, 94; clover, 112–13, *113*, *114*; *Dactylis*, 26; flax, 79–80, *79*; lupin, 99; tobacco, 89–90, *90*; wheat, 141–2, 144–6, *146*
linear and volume, 14–16, *15*, *16*
radial: in flax, 59, 69; in tobacco, 88; and in wheat, *141*, 142
a sensitive index, 10–12, *11*
see also relative rates of change
relative rates of change, 177–82
for: cell surface, *178*; cell-wall material, *178*, *179*; DNA-phosphorus, *178*, *179*; protein-nitrogen, *178*, *179*; RNA-phosphorus, *178*, *179*
one component per unit of another, 179–82
residual dry weight in: clover leaf, *167*, 168–9; wheat leaf, 172, 174–6
Rheo discolor, 106
Richards's function, *see* growth curves
RNA-phosphorus in: clover leaf, *167*, 169–71, *170*; wheat leaf, 172, 175–7, *176*, *177*
runner bud (clover), *see* axillary bud

Scenedesmus costulatus, growth rate, 13
serial reconstruction
and volume estimation, 227–30
Wilhelm His and, 4–5
serradella, 216
see Ornithopus compressus
shoot apex
integrity of, 217
linear and angular measurement, 231–2, *231*
organization of, 214–22
see also shoot-apical systems and apical dome
shoot-apical systems, 4, 56–162
in: cauliflower, 91; clover, 100–3; *Dianella*, 157; eucalyptus, 116–23; fig, 146–54; flax, 56–63; lupin, 97; serradella, 157; sunflower, 160–2; tobacco, 81–6; wheat, 131–5
spike, 193–4, *193*
see also inflorescence
spikelet, 183, *189*, *191*, 192, *193*, *196*, *197*
stamen, 185, *188*, *190*, *191*, 192, 195–7, *195*, *196*, *197*
stipules in: *Acacia*, 34, 160; cauliflower, 94; chrysanthemum, 221, *103*; clover, 101, *103*; fig, *147*, *148*, 149, *149*, *150*, 151–6, *152*, *153*, *154*, *156*, 211; lupin, 97; serradella, *158*
Stylosanthes humilis, 25

sub-apical region, growth rate: in flax, 73, 208; and in wheat, 138–9, 210
sunflower, see *Helianthus annus*

Thaladiantha dubia, 107
tillers
'escape' of bud, 201–3, *203*
growth of, 199–206
see also axillary bud
tobacco, 38, 210, 216
see also Nicotiana tabacum
tomato, 217, 218, 236
transverse projection, 28, *29*
ideal, 32, *32, 33*, 35–6, *36*, 127
in: cauliflower, *96, 215*; clover, 105; eucalyptus, *127*; fig, *155*; flax, *62*, 63, *63*; tobacco, *86, 87*; wheat, 136, *136*
method, 232–3, *233*
transverse section
diagram: for cauliflower, *95*; for eucalyptus, *125, 126*; for fig, *149, 150*; for flax, *62, 80*; for *Mentha, 127*; and for wheat, *191, 200, 201*
photomicrograph: for *Acacia, 160*; for cauliflower, *93*; for clover, *104*; for fig, *148*; for flax, *60*; for tobacco, *85*; and for wheat, *134*
Trifolium repens, 101, 107–8, *108*, 109, 218
Trifolium subterraneum, 41, 100–16, 163–72, 177–82, 208–9, 210, 217, 221

Trigonella foenum-graecum, uni- and tri-foliolate leaves, 214
tri-iodobenzoic acid, 221
Triticum aestivum, 13, *21–4*, 30, *30, 41*, 131–46, 172–7, 177–82, 183–98, 199–206, 209–10, 217, 227–30, *228, 239*
tunica in: cauliflower, 94, *96*; eucalyptus, *120, 124*, 230; flax, *61*, 228–9, *229*; lupin, 100, *100*; wheat, *133*, 228
variability
of leaf length: in clover, 164–5, *165*; and in wheat, 173–4, *173*
of organisms, 3
Vibrio marinus, growth rate, 13
virus, T.M.V., growth rate, 13
volume estimation
for eucalyptus, 118, *120*, 230
large primordia, 230
small primordia, 229–30, *229*
by serial reconstruction, 227–30, *228*
volume growth in: cauliflower, 94, *96*; clover, 110–13, *112*, 168; eucalyptus, 130, *130, 131*; fig, 155, *156*; flax, 75, *76*, 77, 78–9, *78*; lupin, *100*; tobacco, 189–90, *189*; wheat, *138*, 140–1, 195–7, *196, 197*

wheat, see *Triticum aestivum*
Willia anomala, growth rate, 13

yeast, growth rate, 13